I0880806

Principles and Applications of Quantum Computing Using Essential Math

A. Daniel
Amity University, India

M. Arvindhan
Galgotias University, India

Kiranmai Bellam
Prairie View A&M University, USA

N. Krishnaraj
SRM University, India

A volume in the Advances in Computer and Electrical Engineering (ACEE) Book Series

IGI Global
PUBLISHER of TIMELY KNOWLEDGE

Published in the United States of America by
IGI Global
Engineering Science Reference (an imprint of IGI Global)
701 E. Chocolate Avenue
Hershey PA, USA 17033
Tel: 717-533-8845
Fax: 717-533-8661
E-mail: cust@igi-global.com
Web site: http://www.igi-global.com

Copyright © 2023 by IGI Global. All rights reserved. No part of this publication may be reproduced, stored or distributed in any form or by any means, electronic or mechanical, including photocopying, without written permission from the publisher.
Product or company names used in this set are for identification purposes only. Inclusion of the names of the products or companies does not indicate a claim of ownership by IGI Global of the trademark or registered trademark.

Library of Congress Cataloging-in-Publication Data

Names: Daniel, A., 1988- editor. | Arvindhan, M., 1985- editor. | Bellam, Kiranmai, 1982- editor. | Krishnaraj, N., 1984- editor.
Title: Principles and applications of quantum computing using essential math / A. Daniel, M. Arvindhan, Kiranmai Bellam, N. Krishnaraj.
Description: Hershey, PA : Engineering Science Reference, [2023] | Includes bibliographical references and index. | Summary: "The emphasis of the book is on understanding the principles and applications of quantum computing using only essential math-all relevant mathematical concepts are introduced at appropriate places in the text"-- Provided by publisher.
Identifiers: LCCN 2023027718 (print) | LCCN 2023027719 (ebook) | ISBN 9781668475355 (hardcover) | ISBN 9781668475362 (paperback) | ISBN 9781668475379 (ebook)
Subjects: LCSH: Quantum computing--Mathematics.
Classification: LCC QA76.889 .P75 2023 (print) | LCC QA76.889 (ebook) | DDC 006.3/843--dc23/eng/20230907
LC record available at https://lccn.loc.gov/2023027718
LC ebook record available at https://lccn.loc.gov/2023027719

This book is published in the IGI Global book series Advances in Computer and Electrical Engineering (ACEE) (ISSN: 2327-039X; eISSN: 2327-0403)

British Cataloguing in Publication Data
A Cataloguing in Publication record for this book is available from the British Library.

All work contributed to this book is new, previously-unpublished material.
The views expressed in this book are those of the authors, but not necessarily of the publisher.

For electronic access to this publication, please contact: eresources@igi-global.com.

Advances in Computer and Electrical Engineering (ACEE) Book Series

Srikanta Patnaik
SOA University, India

ISSN:2327-039X
EISSN:2327-0403

MISSION

The fields of computer engineering and electrical engineering encompass a broad range of interdisciplinary topics allowing for expansive research developments across multiple fields. Research in these areas continues to develop and become increasingly important as computer and electrical systems have become an integral part of everyday life.

The **Advances in Computer and Electrical Engineering (ACEE) Book Series** aims to publish research on diverse topics pertaining to computer engineering and electrical engineering. **ACEE** encourages scholarly discourse on the latest applications, tools, and methodologies being implemented in the field for the design and development of computer and electrical systems.

COVERAGE

- Computer Architecture
- VLSI Fabrication
- Optical Electronics
- VLSI Design
- Algorithms
- Applied Electromagnetics
- Electrical Power Conversion
- Programming
- Qualitative Methods
- Chip Design

IGI Global is currently accepting manuscripts for publication within this series. To submit a proposal for a volume in this series, please contact our Acquisition Editors at Acquisitions@igi-global.com or visit: http://www.igi-global.com/publish/.

The Advances in Computer and Electrical Engineering (ACEE) Book Series (ISSN 2327-039X) is published by IGI Global, 701 E. Chocolate Avenue, Hershey, PA 17033-1240, USA, www.igi-global.com. This series is composed of titles available for purchase individually; each title is edited to be contextually exclusive from any other title within the series. For pricing and ordering information please visit http://www.igi-global.com/book-series/advances-computer-electrical-engineering/73675. Postmaster: Send all address changes to above address. Copyright © 2023 IGI Global. All rights, including translation in other languages reserved by the publisher. No part of this series may be reproduced or used in any form or by any means – graphics, electronic, or mechanical, including photocopying, recording, taping, or information and retrieval systems – without written permission from the publisher, except for non commercial, educational use, including classroom teaching purposes. The views expressed in this series are those of the authors, but not necessarily of IGI Global.

Titles in this Series

For a list of additional titles in this series, please visit: http://www.igi-global.com/book-series/

Design, Optimization, and Applications of Underwater Acoustic Sensor Networks
Bhumika Gupta (G.B.Pant Institute of Engineering & Technology, India) Kamal Kumar Gola (COER University, India) Brij Mohan Singh (COER University, India) Mridula . (COER University, India) and Rohit Kanauzia (COER University, India)
Engineering Science Reference • © 2023 • 340pp • H/C (ISBN: 9781668456866) • US $270.00

NeutroGeometry, NeutroAlgebra, and SuperHyperAlgebra in Today's World
Florentin Smarandache (University of New Mexico, USA) and Madeline Al Tahan (Abu Dhabi University, UAE)
Engineering Science Reference • © 2023 • 264pp • H/C (ISBN: 9781668447406) • US $270.00

Principles and Theories of Data Mining With RapidMiner
Sarawut Ramjan (Thammasat University, Thailand) and Jirapon Sunkpho (Thammasat University, Thailand)
Engineering Science Reference • © 2023 • 319pp • H/C (ISBN: 9781668447307) • US $255.00

Energy Systems Design for Low-Power Computing
Rathishchandra Ramachandra Gatti (Sahyadri College of Engineering and Management, India) Chandra Singh (Sahyadri College of Engineering and Management, India) Srividya P. (RV College of Engineering, India) and Sandeep Bhat (Sahyadri College of Engineering and Management, India)
Engineering Science Reference • © 2023 • 389pp • H/C (ISBN: 9781668449745) • US $270.00

Artificial Intelligence Applications in Battery Management Systems and Routing Problems in Electric Vehicles
S. Angalaeswari (Vellore Institute of Technology, India) T. Deepa (Vellore Institute of Technology, India) and L. Ashok Kumar (PSG College of Technology, India)
Engineering Science Reference • © 2023 • 342pp • H/C (ISBN: 9781668466315) • US $260.00

IGI Global
PUBLISHER of TIMELY KNOWLEDGE

701 East Chocolate Avenue, Hershey, PA 17033, USA
Tel: 717-533-8845 x100 • Fax: 717-533-8661
E-Mail: cust@igi-global.com • www.igi-global.com

Table of Contents

Detailed Table of Contents

Researchers have focused increasingly on hybrid quantum computing and brain-computer interfaces. Scientists are investigating brain-computer interface technology and applying it to additional fields as neural technology and artificial intelligence evolve. The subject of brain-computer interface has progressed rapidly over the past decades, yet the underlying technologies and novel ideas behind seemingly unconnected systems are rarely summarized from the point of quantum integration. This study describes hybrid quantum computing and brain-computer interface applications, discusses present issues, and suggests future research. Ecologists could employ quantum computers because the statistical approaches they use have proven routes on them. If the hardware, opportunity, and imagination of quantitative ecologists coincide, quantum computing could leapfrog our understanding of complex ecological processes.

In this chapter, the authors will discuss some of the many models and training methods that have been developed in the field of machine learning to address this learning challenge. Models like neural networks and stochastic gradient descent have their own "go-to" training algorithms, each with their own set of supporting terminology and communities of experts. Since the specifics of gate decomposition, compilation, and error correction all depend heavily on the physical implementation of qubits and quantum gates, it has been difficult to design quantum hardware capable of running such algorithms. Therefore, the authors can only provide asymptotic estimations of total execution times. Since developing quantum hardware is so prohibitively expensive, researchers are incentivized to use terms like "superior quantum algorithms" to justify their work. This has given rise to the contentious term "quantum supremacy" to describe experiments that definitively show a difference between classical and quantum levels of computational complexity.

Spintronic systems, which involve manipulating the spin of electrons, have potential for use in quantum computing due to their unique properties. These systems can create spin-based qubits that are more stable and less prone to decoherence than other types, and can also be used to control interactions between qubits. They can also utilize spin-based quantum gates that can perform specific operations faster and more efficiently. Spintronic systems hold promise for the development of powerful and efficient quantum computers.

The aim of this chapter is to develop an understanding of the electronic and optical properties of quantum well systems. These structures, which can be found in semiconductors, confine particles in one dimension and exhibit discrete energy levels that can be calculated using fundamental quantum mechanics. Quantum wells are formed by sandwiching a material like GaAs between two layers of a material with a wider bandgap like AlAs, and can be grown using techniques such as MBE or CVD. The electronic and optical properties of quantum wells can be modified by altering parameters such as potential and well widths. In this chapter, the authors will use the Schrödinger equation to solve for the energy levels of quantum wells and provide a quantum mechanical description of the properties of electrons in these systems. They will also use computer programs to investigate the effects of changing parameters such as potential and well widths on the properties of quantum wells.

Preface

In the rapidly evolving landscape of modern technology, the emergence of quantum computing stands as a beacon of transformative potential. As Editors of *Principles and Applications of Quantum Computing Using Essential Math*, we are delighted to present this reference book, which delves into the fascinating realm of quantum computing, a paradigm poised to revolutionize computation itself.

The era of classical computing has bestowed upon us countless advancements, but certain challenges remained insurmountable due to their inherent complexity. Quantum computing presents a novel approach, harnessing the principles of quantum mechanics to process information in ways that classical computers simply cannot emulate. This book seeks to demystify this intricate subject, guided by the belief that fundamental concepts should be accessible to all, irrespective of their mathematical backgrounds.

We stand at the forefront of a quantum-technology revolution, with both global enterprises and governments investing significant resources into the realization of quantum computing's potential. As we embark on this journey, we witness the emergence of quantum ecosystems, bridging the gap between industry, academia, and startups. The vibrant collaborations fostered within these ecosystems are instrumental in driving the development of quantum technology and the cultivation of a skilled workforce.

In crafting this book, our aim is clear: to provide an accessible and comprehensive resource that guides readers through the principles and applications of quantum computing, employing only essential mathematical foundations. We have meticulously organized each chapter to introduce concepts progressively, ensuring that mathematical explanations are thoughtfully placed to facilitate understanding.

Starting with the basics, our book takes readers through the fundamental principles of quantum computing, introducing quantum bits, gates, circuits, and the intriguing phenomena of entanglement and interference. It delves into quantum algorithms, exploring their applications in areas like machine learning and optimization. Quantum mechanics, a cornerstone of this field, is introduced to build a strong foundation for further exploration.

The book culminates in chapters that embrace the cutting-edge developments in quantum computing, from quantum deep learning to generative adversarial networks. These explorations exemplify the multidisciplinary nature of quantum computing, offering insights into its potential to revolutionize fields such as biology, chemistry, and material science.

Chapter Overview

The opening chapter addresses the intriguing intersection of hybrid quantum computing and brain-computer interface technology. It explores the ongoing integration of these fields and how quantum principles can enhance brain-computer interface applications. The challenges and future prospects of this synergy are discussed, highlighting the potential for revolutionary breakthroughs in various domains.

In chapter 2, we delve into the multifaceted landscape of machine learning models and training methods. With a focus on tackling the learning challenge, we explore neural networks, stochastic gradient descent, and other techniques. The intricate relationship between quantum hardware and algorithm design is unveiled, shedding light on the complex nature of quantum algorithms and their execution times.

Chapter 3 delves into spintronic systems and their role in quantum computing. The chapter introduces the concept of spin-based qubits, their unique stability properties, and the potential for efficient quantum gates. The promise of spintronic systems in revolutionizing the development of powerful and stable quantum computers is explored in depth.

The focus of Chapter 4 is on quantum well systems and their electronic and optical properties. Through the lens of quantum mechanics, the chapter delves into the energy levels and behaviors of particles within quantum wells. Techniques for modifying the properties of quantum wells are explored, highlighting their potential impact on various applications.

In Chapter 5, quantum dots take center stage as intricate semiconductor nanostructures with unique electronic and optical properties. The Schrödinger equation is employed to unravel the energy levels of different quantum dot types. The chapter also utilizes computational tools to investigate the effects of parameter variations on quantum dot properties.

Chapter 6 emphasizes the practical application of quantum computing, illustrating its impact through qubits, coherence time, and gate error rate measurements. The role of quantum computing in data analysis, image processing, and real-time applications is explored. The "noisy intermediate scale quantum" (NISA) concept is introduced, demonstrating its potential in IoT and image processing.

In Chapter 7, the concept of quantum information encoding is unveiled, accompanied by the intriguing concept of quantum entanglement. The implications of entanglement for quantum communication and quantum computing are discussed. Quantum-enabled machine learning and quantum support vector machines (QSVMs) are introduced as innovative applications.

Chapter 8 delves into quantum circuits with statistical properties known as t-designs, shedding light on gradients' behavior in high dimensions. The chapter showcases real-world applications of quantum machine learning across various fields, including drug discovery, financial modeling, image identification, and natural language processing.

The final chapter explores the synergy between emerging technologies, quantum information science, and artificial education systems. It examines the potential crossover between quantum computing and machine learning/artificial intelligence. The chapter delves into recent advancements in both domains and their implications for the future of technology and human civilization.

As editors, we extend our gratitude to the contributors who have crafted each chapter with care and expertise, and to the readers who embark on this journey of exploration. We recognize the immense support from organizations and individuals dedicated to advancing quantum computing, and we acknowledge the pioneers, both within and outside the quantum industry, who are shaping the future of technology.

This book, a culmination of collective effort, seeks to contribute to the growth of quantum computing knowledge, enabling both newcomers and seasoned professionals to navigate this exciting domain. It is our hope that the principles and applications presented within these pages inspire further inquiry, spark innovation, and lay the groundwork for a quantum-powered future.

A. Daniel
Amity University, India

M. Arvindhan
Galgotias University, India

Kiranmai Bellam
Prairie View A&M University, USA

N. Krishnaraj
SRM University, India

Chapter 1
Quantum Metrology:
A Key Ingredient for Advancing Quantum Technologies

Arvindhan Muthusamy
Galgotias University, India

ABSTRACT

Researchers have focused increasingly on hybrid quantum computing and brain-computer interfaces. Scientists are investigating brain-computer interface technology and applying it to additional fields as neural technology and artificial intelligence evolve. The subject of brain-computer interface has progressed rapidly over the past decades, yet the underlying technologies and novel ideas behind seemingly unconnected systems are rarely summarized from the point of quantum integration. This study describes hybrid quantum computing and brain-computer interface applications, discusses present issues, and suggests future research. Ecologists could employ quantum computers because the statistical approaches they use have proven routes on them. If the hardware, opportunity, and imagination of quantitative ecologists coincide, quantum computing could leapfrog our understanding of complex ecological processes.

DOI: 10.4018/978-1-6684-7535-5.ch001

Copyright © 2023, IGI Global. Copying or distributing in print or electronic forms without written permission of IGI Global is prohibited.

1. INTRODUCTION

The modern machines we use today are the result of a long line of conceptual breakthroughs followed by incremental hardware improvements. In a similar vein, they imply that the methods currently in use to represent logical qubits, such as a spin transmon or a gateman, may be effective learning devices, but are ultimately unscalable. They imply that an alternative strategy must be developed that can be scaled. A novel hardware-stable method of representing logical qubits has been developed, they say. According to the claims, the apparatus can bring about a state of matter with Majorana zero modes, a class of fermions. They also claim to have evidence of devices meeting this description, reporting low enough disorder to pass the topological gap procedure. They consider this a breakthrough on the road to developing not just a quantum computer but a quantum supercomputer. Microsoft also revealed a new metric for measuring the efficacy of a quantum supercomputer, which it calls "reliable quantum operations per second" (rQOPS). This value indicates the number of reliably executed operations that can be performed by the computer in a certain period. According to their criteria, a machine's QOPS must be greater than 1 million to be considered a quantum supercomputer. They point out that machines with such capabilities may achieve a billion QOPS, which is truly impressive().

2. THE POWER OF QUANTUM TECHNOLOGIES: A DEEP DIVE INTO QUANTUM METROLOGY

As a new domain of quantum technologies, quantum metrology has been getting a lot of attention from academics and professionals in related fields. Quantum metrology has emerged as an essential component for advancing quantum technologies as the world works towards harnessing the potential of quantum mechanics for practical applications. The success of quantum technologies like quantum computing, quantum communication, and quantum sensing depends on the development of high-precision measurement instruments and procedures, which are the focus of this cutting-edge field.

The possibility that quantum metrology would radically alter how we measure and make sense of the world is a major factor in the field's rising profile. Quantum metrology, which takes advantage of quantum phenomena like superposition and entanglement, has the potential to achieve hitherto unattainable levels of precision and accuracy in measurement. As a result, the efficiency of many quantum technologies can be greatly improved, opening the door to new scientific and technological breakthroughs().

For instance, to fully use the promise of quantum computers, extremely precise and stable quantum bits (qubits) must be developed. By giving the means to precisely characterize and manipulate qubits, quantum metrology can pave the way for the development of more stable and efficient quantum computers. Quantum communication relies on accurate measurement and control of quantum states for safe and efficient data transfer. Due to the limits of classical communication systems, there is a great need for sophisticated measurement techniques that quantum metrology can provide.

The use of quantum systems to detect and measure physical quantities like magnetic fields, gravitational forces, and temperature is known as quantum sensing, and quantum metrology has the potential to revolutionize this discipline. Researchers can create quantum sensors that are more precise and accurate than their conventional analogues by making use of the increased sensitivity and resolution provided by quantum metrology. This has a wide range of potential uses, from monitoring brain activity in medical diagnostics to detecting subtle shifts in the Earth's gravitational field for geophysical investigation().

The benefits of quantum metrology are not confined to the quantum world. The methods and instruments created in this domain can have far-reaching effects on traditional technologies, especially in settings where extremely precise measurements are required. In the case of the Global Positioning System (GPS), which uses accurate timekeeping for navigation, the enhanced atomic clocks produced utilizing principles of quantum metrology can be of great use. In a similar vein, quantum metrology-based improved imaging approaches can improve the diagnostic accuracy and patient care provided by medical imaging systems by increasing their resolution and sensitivity (Lin et al., 2021).

Despite quantum metrology's promising future, some obstacles must be overcome before it can be put to good use. Creating quantum systems that are both scalable and feasible for real-world use is a huge challenge. In addition, the high levels of precision and accuracy needed in quantum metrology are difficult to maintain due to quantum systems' sensitivity to external noise and decoherence. Researchers are attempting to improve the reliability and functionality of quantum systems by creating new materials, methodologies, and error correction strategies (Childs et al., 2002).

Figure 1. Difference between quantum vs. classical computing

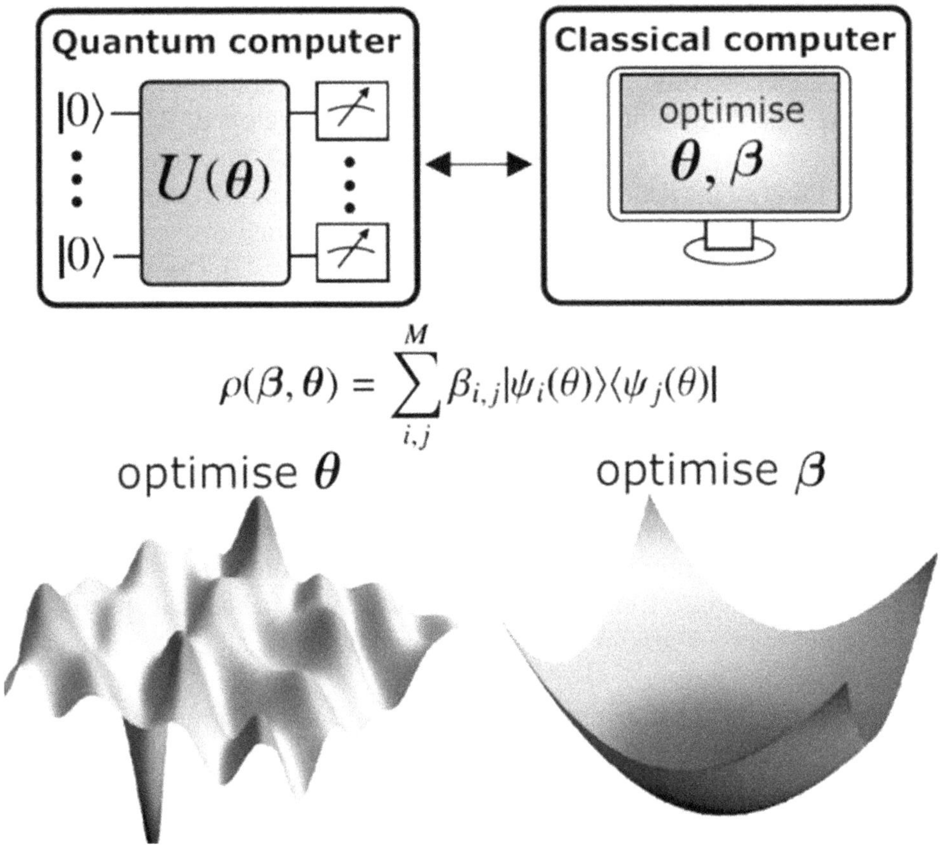

In conclusion, quantum metrology will be a critical component in developing quantum technologies and influencing the direction of the scientific and technological communities. Quantum metrology provides the means to attain unprecedented levels of precision and accuracy in measurement, which can be used to realize the full promise of quantum computing, communication, and sensing as well as to improve traditional technology Figure 1. We can anticipate a new era of invention and discovery that will radically alter our perspective of the world as scientists devote more time and energy to studying the foundations of quantum metrology.

Figure 2. Semiconductor region of quantum dots

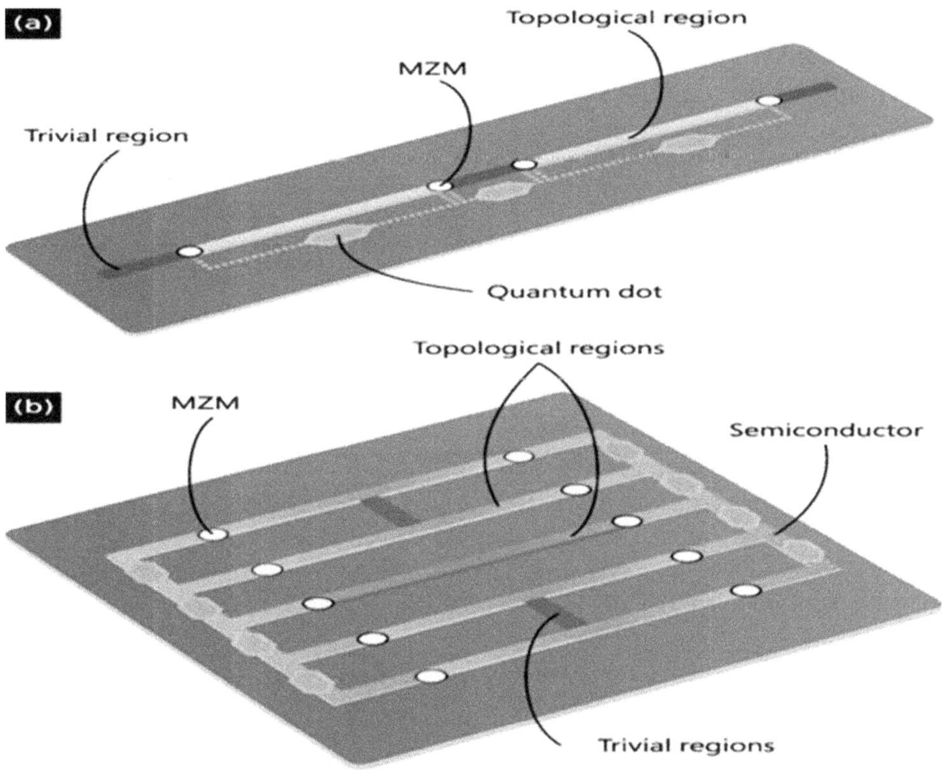

3. VARIATIONAL QUANTUM ALGORITHMS

One of the most pressing technological questions that have arisen since the advent of quantum computing as a viable alternative to classical computing for problems of high computational complexity like large-scale linear algebra and discrete optimization is this: how can the current crop of NISQ (Noisy Intermediate Scale Quantum) devices be used to their full potential to achieve quantum advantage? With these limitations in mind, Variational Quantum Algorithms have emerged as a viable answer to this problem. Similar to how neural networks are trained, these algorithms take the idea of training quantum computers and apply it to discovering the optimal parameters of the model to minimize/maximize some objective function associated with that model. To make use of NISQ hardware, there must be some part of an algorithm that can be executed by a quantum circuit. These quantum-classical algorithms of the near future are called Variational Quantum algorithms, and the Variational Quantum Circuit explains the quantum subroutine

(Hallgren et al., 2000). Measurement statistics are generated from the input data (or the state preparation) using this quantum circuit. It's also limited by some variables. The optimizer uses a conventional outer loop to update the parameters while quantum resources are used to process the measurements and the cost functions Figure 2.

4. BRIEF OVERVIEW OF VQA

Inspiring Quantum Variational Algorithms is the Variational Principle of Quantum Mechanics, which will be discussed in more depth shortly. These algorithms can estimate solutions to a problem and are modular circuit-based, problem statement agnostic. The flexibility of VQAs translates to a wide range of algorithmic structures that can range in complexity, although all variational algorithms share the same fundamental framework Fig.3.

Figure 3. Quantum variational algorithm

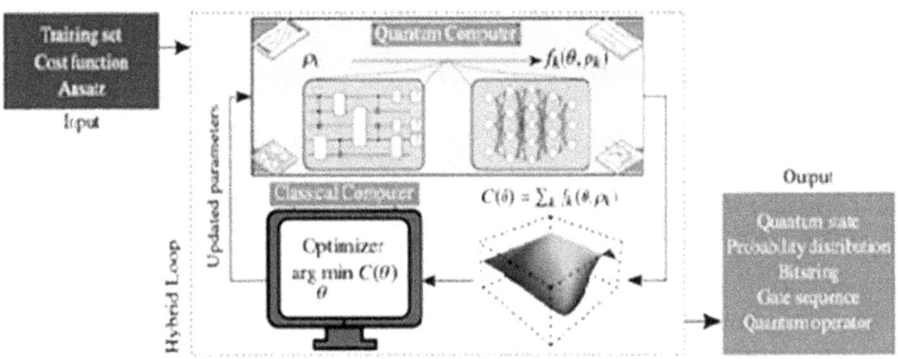

The problem is initialized by setting the quantum computer to its default state, |00..0>, and then applying the transformation of a non-parameterized unitary reference operator Ur to arrive at the reference state, |R>.

The circuit's "Ansatz," or design, is not always a constant depending on the algorithm being used. Facilitates our transition from [00..0] to [Y(x)]. A variational operator of the type Uv(x), where Uv(x) is a set of sharable parameters, facilitates the transformation. The aggregate concept of the Ansatz operator might be thought of as:

$$E(Y) = \langle Y|H|Y\rangle = \left\langle Y \left| \sum_{i \in Spec(H)} \left(E_i |Y_i Y_i|\right) \right| Y \right\rangle$$
$$= \sum_{i \in Spec(H)} E_i Y \; Y_i Y_i \; Y = \sum_{i \in Spec(H)} E_i |Y \; Y_i|^2 \geq E0\langle Y|Y\rangle = E_Y \quad (1)$$

The problem is initialized by setting the quantum computer to its default state, |00..0>, and then applying the transformation of a non-parameterized unitary reference operator Ur to arrive at the reference state, |R>.

The circuit's "Ansatz," or design, is not always a constant depending on the algorithm being used. Facilitates our transition from () to [Y(x)]. A variational operator of the type Uv(x), where Uv(x) is a set of sharable parameters, facilitates the transformation. The aggregate concept of the Ansatz operator might be thought of as:

$$U_A(x) = U_V(x)U_R \quad (2)$$

Cost function: An objective function tailored to the unique problem at hand is needed to systematically maximize or minimize the target metric. We represent it with the quantum system's linear combination of Pauli operators, denoted by C(t).

Method of instruction: This subroutine iteratively modifies the circuit parameters based on the information gained from the output measurements of the quantum circuit, intending to reach an optimal solution. If a less-than-ideal solution is already known, it can be used as a starting point for a bootstrapped optimization procedure (Childs et al., 2001).

The Principle of Variation

One of the most fundamental principles in Quantum Mechanics is the variational principle. It aids in estimating the energies of excited states and the system's ground state. It helps when an analytical solution to the Schrodinger equation can't be found, and its ramifications are far-reaching. Wavefunctions of molecular orbitals, for instance, can be estimated using the variational approach (Andrew, 2003).

Simply put, this technique entails picking an unnormalized trial wavefunction Y(x) that is sensitive to the variational parameters x. We then search for x-values where the expected energy of the system is the smallest. The resultant wavefunction approximates the ground state wavefunction at these parameter values, and the expectation value provides an upper bound on the ground state energy (Andrew, 2007).

The Hamiltonian is a quantum observable that represents energy in quantum mechanics. If we know the maximum value of the ground-state energy, what use is that information to us? Why? Because the maximum energy of a physical system provides a reliable approximation of the true value. The concept of ground state energy is fundamental to quantum chemistry since it allows us to put a numerical value on the subtleties of molecule features like binding energy, molecular routes, etc. Our ability to accurately deduce the succeeding, more complex attributes depends on how near our approximation for ground state energy is to the genuine value. Using quantum computers is a huge assistance in these situations since they guarantee a high approximation ratio.

Let's Get Into the Variational Principle's Formalization

Given a Hilbert space and Hermitian operators over it, dubbed a Hamiltonian H, we examine the independent spectrum of H on a basis formed by |Yi> (The eigenvectors, where Yj|Yi> = 1 if i=j and 0 otherwise) using Finite The term discrete The spectrum Theorem without considering the difficulties that occur with continuous spectra.

In this case, the equation for H's defining feature,

$$H|Y_i = E_i|Y_i \tag{3}$$

wherein |Yi> is the appropriate eigenstate and Ei is the i-th amplitude (either in person or the i-th energy level), wherein i is a number in the N-dimensional universe of states. Similarly, H's monochromatic decomposition looks like this:

$$H = \sum_{i \in N} E_i \left| Y_i \, Y_i \right| \tag{3}$$

The expectation value of H, Y|H|Y> = E(Y), can be used to determine the energy value for a specified state |Y>. The expression Yi|H|Yi> = E(Yi) holds if and only if the state under consideration is an eigenstate.

The ground state energy E0 is calculated for the eigenstate corresponding to the lowest eigenvalue.

5. THE ANSATZ

The term "ansatz" is used in the context of quantum computing to refer to the process of making a best estimate as to the value/form/solution of a function whose exact value/form/solution is unknown. They have been employed in well-known algorithms like VQEs, QAOAs, etc., and have direct importance in efficiently performing Variational Quantum Algorithms on NISQ Hardware. As a trial (beginning) point, an ansatz—also called the state-preparation circuit/variational form—is a subroutine built by parameterized gates working on specific wires in a quantum circuit. In the same way that artificial neural networks are parameterized mathematical models whose goal is to approximate real-world, high-dimensional, unknown functions, this model is analogous to the architecture of such networks (Figure 4).

Let's circle back to the original impetus for ansatz. We set out to efficiently search all of Hilbert's space for the variational solution (represented by the smallest eigenvalue) to the modelled problem. The use of parameterized circuits provides a viable solution. If the input state is a vector of a polynomial number of circuit parameters, then the output state is a tunable unitary operation on n qubits, U(x). Consider a hypothetical circuit in which a single RY gate is present. Only one parameter, the rotation angle x0, is applied to it. We can explore a large variety of quantum states with this rather straightforward setup. By incorporating extra gates like RX and RZ, we can increase the number of quantum states we can traverse without resorting to full Hilbert space exploration. With the help of an ansatz, we may confine the variational search to a smaller, more relevant subspace. Finally, the state that will be used to apply the variational principle will be prepared through an ansatz (Andrew, 2008).

To apply the relevant cost function to the quantum circuit's incoming states, we encode the parameters x in a unitary U(x). U(x) might be written as the sum of L unitary circuits that are each applied to x in turn.

$$U(x) = U_L(x_L) \dots U_2(x_2)U_1(x_1) \tag{3}$$

Getting back to our neural network comparison. Each neuron layer in a neural network is transformed by a non-linear scalar function into the next neuron layer in the network. These are all parts of its hidden internal layers, and they help it generalize the data quite well. We can now see, using ansatzes as well, that this structure is strikingly similar to the variational quantum circuit model. Our input states are transformed because the gates in the quantum circuit perform a linear unitary transformation on each layer. One minor distinction is that only the entities corresponding to the adjustable parameters of the gates in a quantum circuit's transformation matrix act as parameters. In this regard, variational quantum circuits can be understood as a highly effective parameterization of high-dimensional linear transformation.

Figure 4. Unitary circuits acting on the input state

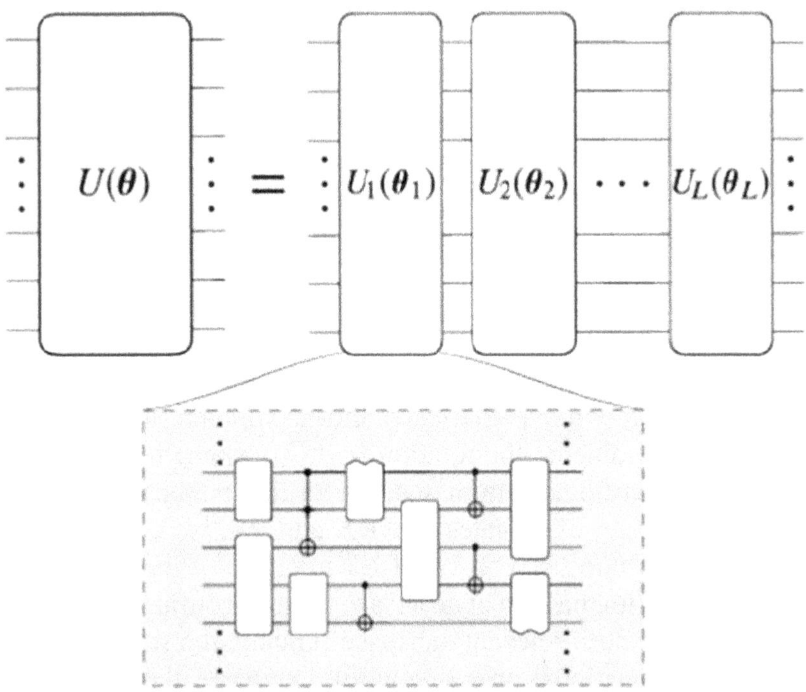

Hardware Efficient Ansatz (HEA), Unitary Coupled Ansatz (UC), Quantum alternating operator ansatz, etc. are only a few examples of the many types of general ansatz architectures that have been presented in the literature.

The effectiveness of a VQA is determined by whether or not its eigenstate can be attained in terms of the least eigenvalue. This depends heavily on the structure of the ansatz and how effectively the parameters have been tailored to the given VQA. In the same way that model architectures in Machine Learning adhere to the no-free-lunch theorem, the ansatz choices for variational algorithms are context-dependent and no "best" ansatz can be used across the board. There are 'problem-inspired' and 'problem-agnostic' ansatzes, the latter of which can be employed even if no specific information about the problem is known. The idea is to reduce the number of gates (and parameters), making it more robust and providing us more leeway in terms of optimization, while still covering a practical amount of space (Ignacio, 1995; Isaac, 1997).

6. QUANTUM CHEMISTRY

Calculating molecular energies, capturing intricate interactions, and optimizing molecular structures of highly complicated molecules are all examples of the types of challenges in quantum chemistry that VQAs have been applied to. Ground state energy estimates and molecular property simulations are the goals of transferring the electronic structure problem onto a quantum circuit and using VQAs. Potential applications include catalyst optimization, material design, and drug discovery (Figure 5).

Figure 5. Variation in quantum algorithm

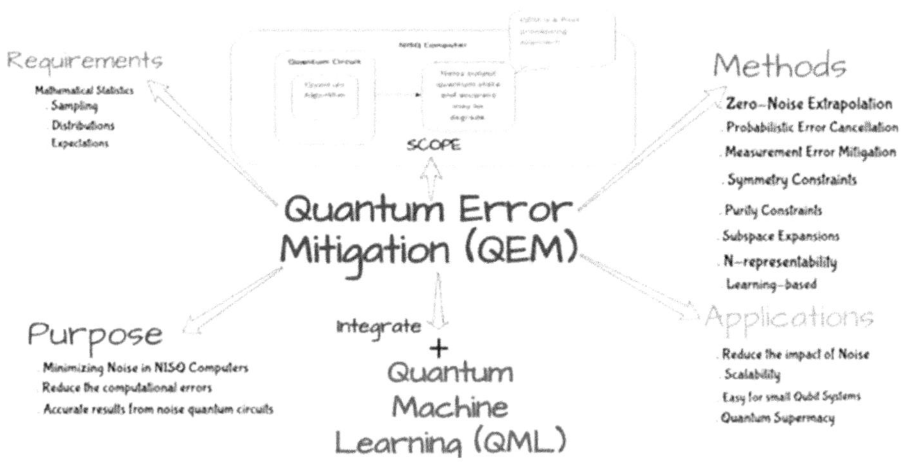

6.1 Optimization

As we'll see in a future post, the QAOA is effective in optimizing combinatorial issues, and VQAs can be applied to optimization problems in a wide range of fields. The QAOA has been used to find effective routes with the least distance travelled in several combinatorial optimization problems, including the travelling salesman problem (TSP). Portfolio optimization, logistics planning, and supply chain management are three other areas where VQAs have been put to use since arriving at the best possible solutions can have a major impact on productivity and usefulness (Cleve et al., 1999).

6.2 Machine Learning

Projects involving machine learning such as data classification and pattern recognition have been investigated using VQAs. VQAs show potential for enhancing the management of complicated datasets by utilizing quantum circuits as feature maps or kernels. Machine learning methods that use VQAs include Quantum Generative Adversarial Networks (QGANs) and Variational Quantum Support Vector Machines (VQSVMs). These use cases show how VQAs can be put to good use in the real world to improve machine learning by implementing quantum computing ideas (Graham, 2006; James, 1965).

7. QUANTUM ERRORS AND NOISE

The two main obstacles to improved and more precise calculation results in quantum computing are noise and errors. Both of these topics drive the need for additional study into Quantum Computing and Quantum Machine Learning models. The difference between each of these is seen in the illustration (Figure 6).

Figure 6. Quantum error mitigation: QEM

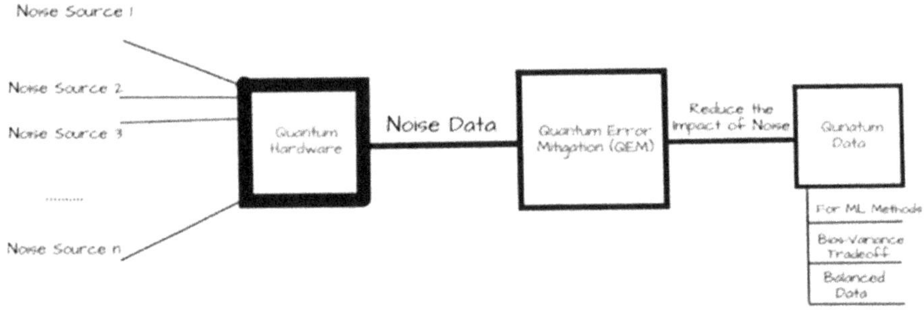

The term "noise" refers to a collection of influences that reduce the precision of NISQ devices' calculations. The quantum information stored in a quantum computer's idle qubit can be easily eroded by noise from a variety of sources. Quantum logical operations on qubits are another potential source of mistakes in quantum data. The final quantum configuration of the quantum computer does not correspond to the precise condition that was anticipated. The quantum information is either randomly generated or completely deleted to prevent these noise-causing issues. The term " Decoherence" describes this physicist term. In a quantum computer, noise can occur. Errors in quantum computation are unavoidable because it is challenging to adequately isolate the qubits from the effects of external noise (Figure 7).

Figure 7. NISQ in quantum computers

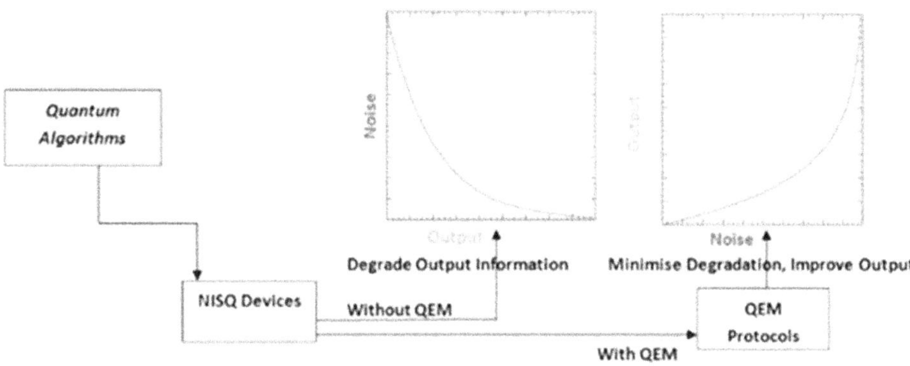

8. QUANTUM ERROR VS. QUANTUM NOISE

In Quantum Hardware, errors can arise in the qubits during the storing, transforming, or transporting of information, while noise can occur in the calculations, state preparation, or final measurement. Please be aware that some quantity of background noise is created by each quantum gate operation.

8.1 Types of Errors and Quantum Computers

Qubits are affected by Quantum Noise, which causes mistakes in the basis states. Bit-flip errors and phase-flip mistakes are the two most common types of flip faults in quantum computing. When the qubit's state transitions from |0> to |1> or vice versa, bit-flip errors can occur. One other name for this is X-errors. While |0> always remains |0>, phase-flip errors are possible during |1> to transitions to -|1>. Both can have a multiplicative effect on quantum system faults (Coppersmith, 1994).

Quantum computers are vulnerable to several different types of defects, including gate flaws, decoherence errors, measurement errors, and cross-talk problems. The forthcoming article Quantum Error Correction delves deeper into these mistakes (Figure 8).

Figure 8. Noise ratio in quantum error mitigation

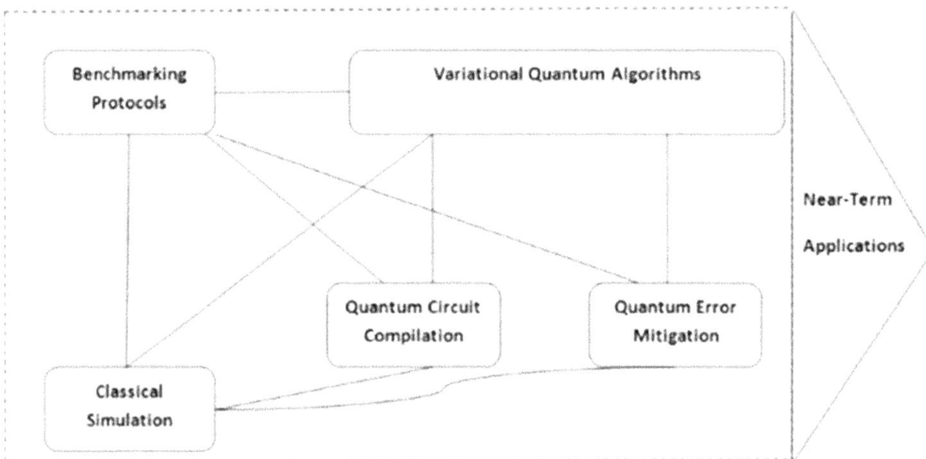

9. QUANTUM HARDWARE — BIAS — NOISE — QEM

Making a ruckus with quantum processes requires dedicated quantum hardware. For both ideal and nonideal quantum systems to tolerate background noise. Zero bias and a complete absence of noise would characterize ideal hardware, but the reality of flawed hardware necessitates the use of QEM to address finite bias via its protocols. QEM can evaluate precise outcomes from noisy quantum circuits by reducing computational mistakes (Figure 9).

10. QEM IS NISQ TECHNIQUE

Quantum Error Mitigation (QEM), Quantum Error Mitigation (QEM), Quantum Circuit Compilation (QCC), and Benchmarking protocols are all examples of NISQ Techniques that are working together. Below the image is some text explaining the connection between the various NISQ methods (Crépeau et al., 2002; Cross et al., 2007; Thomas, 2001). The combination of these methods allows for the successful completion of a broad variety of complicated computing tasks in fields as diverse as mathematics, machine learning, cryptography, drug discovery, optimization, and many more (Figure 10).

While QEC has the potential to make noise-tolerant Quantum Computing a reality, it is currently outside the capabilities of most Quantum processes. QEM enters the picture to solve this problem.

Near-term quantum processes are the primary focus of QEM. QEM gives us a practical option to reduce faults in upcoming operations and a seamless transition from current quantum hardware to future fault-tolerant quantum computers.

Figure 9. NISQ device with QEM

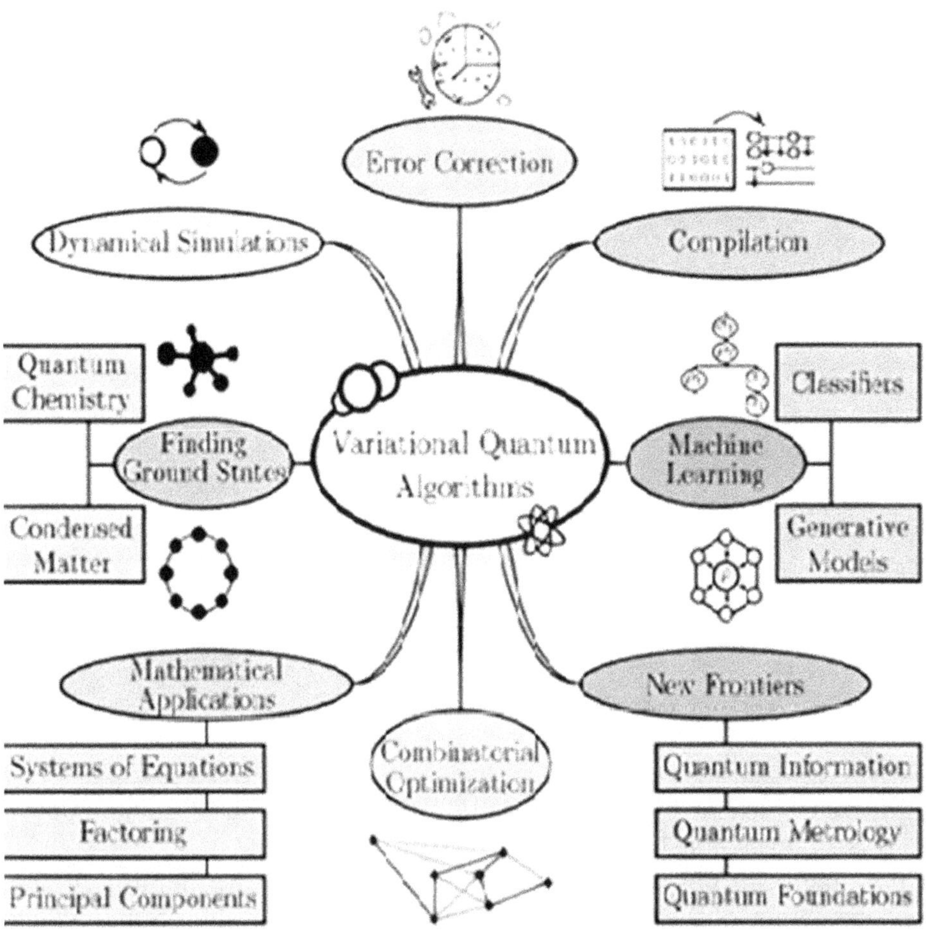

Figure 10. Protocol variation for quantum algorithm

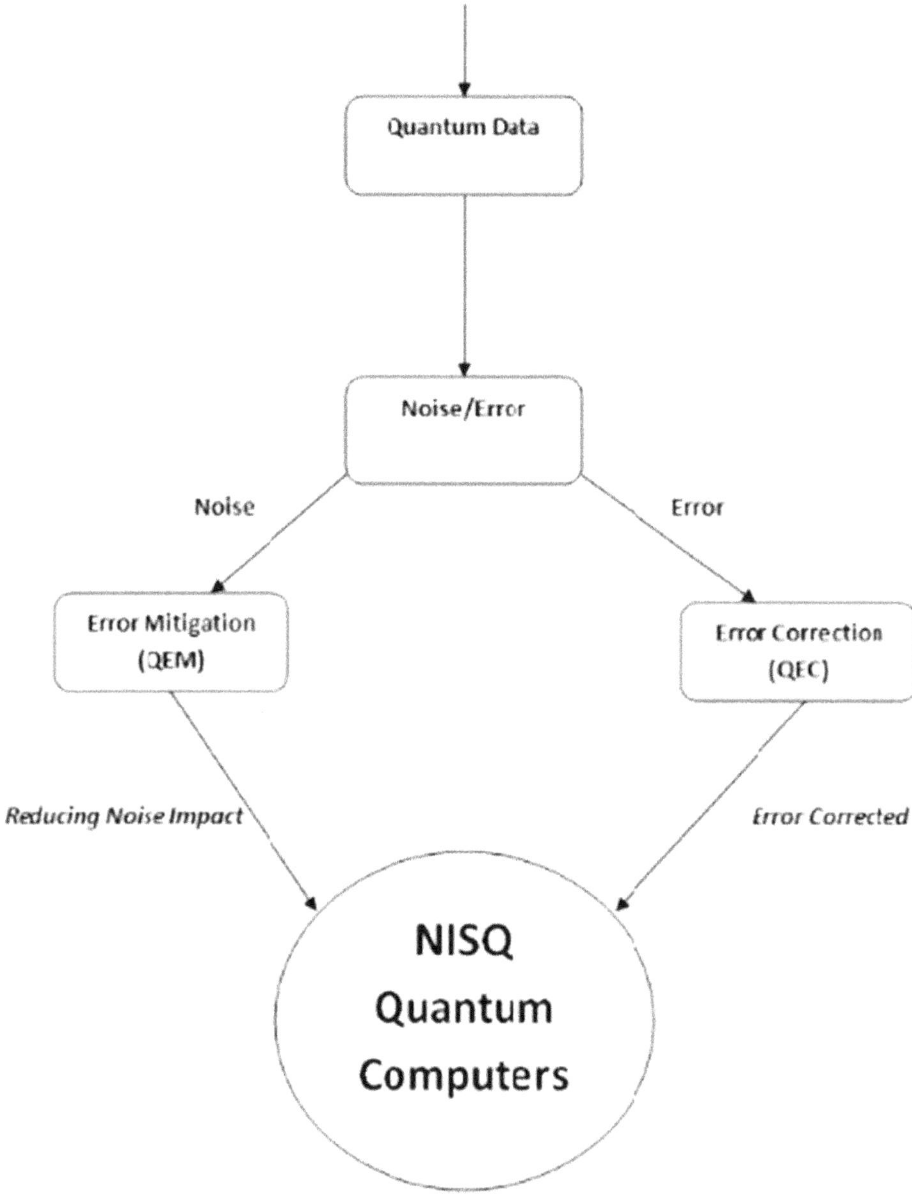

When it comes to minimizing the effects of Noise in NISQ computations, QEM is an invaluable tool.

For many purposes, Quantum Neural Networks are helpful in the implementation of Machine Learning /Deep Learning algorithms, which are seen as the instrument for dealing with QEM because of their ability to mitigate noise-based and decoherence-related problems. The use of a deep learning-based approach to lower readout errors on quantum hardware is demonstrated in one of the studies (Feynman et al., 2010; Hallgren et al., 2003; Hardy & Wright, 1979; Hillery et al., 1999; Zeng et al., 2017).

Because of its ability to reduce noise, quantum electromagnetism (QEM) has close ties to artificial intelligence (AI), machine learning (ML), and deep learning (DL). QEM protocols have substantial applications in the Learning domain, specifically in the state preparation and final measurement stages of quantum algorithms.

Overfitting is a problem in the Machine Learning field, and noise and decoherence may help solve this problem during training. Many noise-reduction methods can be applied in conjunction with Neural Networks.

Applications of Quantum Computing Ranked From Most Important to Least Important

Material Science

To begin, the potential for quantum computing to simulate quantum systems has applications in fields as diverse as materials science and drug discovery.

There are dedicated quantum computing research teams at companies like Google, IBM, Microsoft, and Intel, and businesses like Airbus, Volkswagen, and JPMorgan Chase are actively looking for answers to some of their most pressing issues.

To stay ahead of the curve in this new industry, several huge corporations are beginning to experiment with quantum computing. Smaller businesses and startups are also participating.

As the complexity of molecular interactions increases, the reliability of computations performed on conventional computers decreases. IBM Institute for Business Value's report Exploring quantum use cases for chemicals and Petroleum cites "Developing chemical products, including catalysts and surfactants" as an application where the use of quantum computers speeds up the development of new chemical methods and materials (Gao et al., 2021).

The article continues, citing a 2017 Nature cover story, to discuss how IBM's open-source quantum computers could be used to model lithium hydride (LiH) and beryllium hydride (BeH2) in the chemical and petroleum industries. The paper suggested that similar hybrid approaches could be utilized shortly to create novel catalysts for decreasing emissions or surfactants to enhance subsurface recovery.

Here are the top five uses for quantum computers.

Quantum Computing in Finance

The authors conclude that "a marked improvement in computing time requiring a smaller memory footprint has been observed using quantum data processing techniques, clearing the way for their application in real-world applications in the valuation of derivatives" after discussing two experiments on effective derivatives computation and the parties' reductions anticipation. The selected problem was implemented on a quantum computer and tested in realistic settings. The results produced with a 50-qubit quantum processor are on par with those generated in production. Based on our forecasts, we expect that this performance will be surpassed by 300 qubits by 2024 when this power will be commercially available (Wolpaw, 2007).

Quantum Computing Application in Machine Learning (ML)

Part of the piece on the blog acknowledged that quantum computers are expected to offer exponential improvements over classical systems for certain problems, but that to realize all of their capabilities, study participants must first build up the total amount of qubits as well as enhance quantum error correction. McClean and Huang concluded that the current research demonstrated the first increasing advantage of quantum machine learning (QML) when the number of data points extracted from the quantum state was limited, and that this type of quantum learning takes advantage of is insurmountable even by limitless classical computing resources (Leskowitz & Mueller, 2004).

Quantum Computing Applied to Natural Language Processing (NLP)

At Merck, we're dedicated to finding new ways to use quantum computing's peculiarities to make ground-breaking discoveries. Our recently revealed work in QNLP with academics from TU Munich demonstrates that QNLP techniques can produce results equivalent, even at this early level, to conventional classical methods when applied to binary classification tasks for sentences. Developments in the underlying infrastructure for quantum computing are required before these methods may be used in a commercial setting. QNLP's methodology paves the way for explainable AI, which in turn leads to more accurate, accountable cognitive ability, becoming especially important in the field of medicine.

Since quantum computers are thought to be capable of completing difficult optimization tasks at a considerably faster rate than classical computers, they may find useful applications in areas like logistics and manufacturing. Results from the pilot suggest that further development of the system has the potential to enhance the priority mix of people, weight, and volume loaded on crowded flights. It will also enable more effective process management by cutting down on analysis and optimization times (Kontsevich & Manin, 1994).

11. CONCLUSION

Quantum computing is still in its infancy, but it might have far-reaching consequences. This developing technology has the potential to revolutionize computation and have far-reaching consequences for many areas of our life, from weather forecasting to online security. The future is not merely a possibility in quantum mechanics; it is a likelihood. The role of noise in quantum computing has been investigated. In subsequent sections, we will cover QEM protocols and more complex noise types that QC can manage. This article provided a deeper understanding of QEM and its close ties to the field of artificial intelligence and machine learning from a research standpoint.

REFERENCES

Andrew, M. (2003). Childs, Richard Cleve, Enrico Deotto, Edward Farhi, Sam Gutmann, and Daniel A. Spielman. Exponential algorithmic speedup by a quantum walk. In *Proceedings of STOC '03,* (pp. 59–68.). UMD.

Andrew, M. (2007). Childs, Andrew J. Landahl, and Pablo A. Parrilo. Improved quantum algorithms for the ordered search problem via semidefinite programming. *Physical Review A*, *75*(3), 032335. doi:10.1103/PhysRevA.75.032335

Andrew, M. (2008). Childs and Troy Lee. Optimal quantum adversary lower bounds for ordered search. *Lecture Notes in Computer Science*, *5125*, 869–880. doi:10.1007/978-3-540-70575-8_71

Cai, Z., Babbush, R., Benjamin, S., Huggins, W., Li, Y., McClean, J., & O'Brien. (2022). *Quantum Error mitigation.* Cornell University.

Childs, A., Farhi, E., Goldstone, J., & Gutmann, S. (2002). Finding cliques by quantum adiabatic evolution. *Quantum Information & Computation*, *2*(181), 181–191. doi:10.26421/QIC2.3-1

Childs, A., Farhi, E., & Preskill, J. (2001). Robustness of adiabatic quantum computation. *Physical Review A*, *65*(1), 012322. doi:10.1103/PhysRevA.65.012322

CleveR. (1999). *An introduction to quantum complexity theory*. arXiv:quant-ph/9906111v1.

Cleve, R., Gottesman, D., & Lo, H.-K. (1999). How to share a quantum secret. *Physical Review Letters*, *83*(3), 648–651. doi:10.1103/PhysRevLett.83.648

Coppersmith, D. (1994). *An approximate Fourier transform is useful in quantum factoring. Research Report RC 19642*. IBM.

Crépeau, C., Gottesman, D., & Smith, A. (2002). Secure multi-party quantum computation. *Proceedings of STOC '02*, 643–652. 10.1145/509907.510000

Cross, A., DiVincenzo, D. P., & Terhal, B. (2007). A comparative code study for quantum fault tolerance. arXiv:quant-ph/0711.1556v1.

Feynman, R. P., Hibbs, A. R., & Styer, D. F. (2010). *Quantum mechanics and path integrals*. Courier Corporation.

Gao, X., Wang, Y., Chen, X., & Gao, S. (2021). Interface, interaction, and intelligence in generalized brain–computer interfaces. *Trends in Cognitive Sciences*, *25*(8), 671–684. doi:10.1016/j.tics.2021.04.003 PMID:34116918

Graham, P. (2006, April). Collins. Computing with quantum knots. *Scientific American*, *294*(4), 56–63. doi:10.1038cientificamerican0406-56 PMID:16596880

Hallgren, S., Russell, A., & Ta-Shma, A. (2000). Normal subgroup reconstruction and quantum computing using group representations. In *Proceedings of STOC '00*, (pp. 627–635). ACM. 10.1145/335305.335392

Hallgren, S., Russell, A., & Ta-Shma, A. (2003). The hidden subgroup problem and quantum computation using group representations. *SIAM Journal on Computing*, *32*(4), 916–934. doi:10.1137/S009753970139450X

Hardy, G. H., & Wright, E. M. (1979). *An Introduction to the Theory of Numbers*. Oxford University Press.

Hillery, M., Buzek, V., & Berthiaume, A. (1999). Quantum secret sharing. *Physical Review A, 59*(3), 1829–1834. doi:10.1103/PhysRevA.59.1829

Huang, H., Xu, X., Guo, C., Tian, G., Wei, S., Sun, X., Bao, W., & Long, G. (2021). *Quantum Physics.* Cornell University.

Ignacio, C. & Zoller, P. (1995). Quantum computations with cold trapped ions. *Physical Review Letters, 74,* 4091-4094. doi:10.1103/PhysRevLett.74.4091 PMID:10058410

Isaac, L. (1997). Chuang and Michael Nielsen. Prescription for experimental determination of the dynamics of a quantum black box. *Journal of Modern Optics*, *44*, 2567–2573.

James, W. (1965). Cooley and John W. Tukey. An algorithm for the machine calculation of complex Fourier series. *Mathematics of Computation*, *19*(90), 297–301. doi:10.1090/S0025-5718-1965-0178586-1

Kim, J., Oh, B., Chong, Y., Hwang, E., & Park, D. (2021). *Quantum readout error mmitigation via deep learning.* Cornell University.

Kontsevich, M., & Manin, Y. (1994). Gromov-Witten classes, quantum cohomology, and enumerative geometry. *Communications in Mathematical Physics*, *164*(3), 525–562. doi:10.1007/BF02101490

Leskowitz, G. M., & Mueller, L. J. (2004). State interrogation in nuclear magnetic resonance quantum-information processing. *Physical Review A*, *69*(5), 052302. doi:10.1103/PhysRevA.69.052302

Lin, J., Wallman, J. J., Hincks, I., & Laflamme, R. (2021). Independent state and measurement characterization for quantum computers. In Physical Review Research, 3(3). American Physical Society (APS). doi:10.1103/PhysRevResearch.3.033285

Thomas, H. (2001). *Introduction to Algorithms.* MIT Press.

Wolpaw, J. R. (2007, October). Brain-computer interfaces (BCIs) for communication and control. In *Proceedings of the 9th international ACM SIGACCESS conference on Computers and Accessibility* (pp. 1-2). 10.1145/1296843.1296845

Zeng, Y., Xu, P., He, X., Liu, Y., Liu, M., Wang, J., Papoular, D. J., Shlyapnikov, G. V., & Zhan, M. (2017). Entangling two individual atoms of different isotopes via Rydberg blockade. *Physical Review Letters*, *119*(16), 160502. doi:10.1103/PhysRevLett.119.160502 PMID:29099205

Chapter 2
Exploring Models, Training Methods, and Quantum Supremacy in Machine Learning and Quantum Computing

Arvindhan Muthusamy
Galgotias University, India

ABSTRACT

In this chapter, the authors will discuss some of the many models and training methods that have been developed in the field of machine learning to address this learning challenge. Models like neural networks and stochastic gradient descent have their own "go-to" training algorithms, each with their own set of supporting terminology and communities of experts. Since the specifics of gate decomposition, compilation, and error correction all depend heavily on the physical implementation of qubits and quantum gates, it has been difficult to design quantum hardware capable of running such algorithms. Therefore, the authors can only provide asymptotic estimations of total execution times. Since developing quantum hardware is so prohibitively expensive, researchers are incentivized to use terms like "superior quantum algorithms" to justify their work. This has given rise to the contentious term "quantum supremacy" to describe experiments that definitively show a difference between classical and quantum levels of computational complexity.

DOI: 10.4018/978-1-6684-7535-5.ch002

Copyright © 2023, IGI Global. Copying or distributing in print or electronic forms without written permission of IGI Global is prohibited.

Models, Training , Quantum Supremacy in ML, Quantum Computing

1. INTRODUCTION

The term “quantum advantage” is used to describe the unrealized potential of quantum computers to outperform classical computers. It boils down to a contest between quantum and traditional computers in terms of raw computational capability. Given the significance of the idea, it is not unusual for a corporation or university to publish a scientific publication proclaiming quantum superiority in a certain subfield of quantum computing every once in a while. In this post, we look at a new research study published by IBM that shows how they employed error mitigation to outperform a conventional supercomputer using the most up-to-date approximation techniques. The largest quantum circuits yet executed on a quantum computer were produced by this experiment. Quantum computing is getting close to having a quantum advantage, but we may not be there yet (Nielsen & Chuang, 2010).

1.1 Key Areas of Research for Quantum Advantage

According to IBM’s definition, a quantum advantage occurs when a quantum algorithm significantly outperforms the best classical algorithm in terms of runtime for practical applications. It also explains why no classical algorithm could effectively simulate quantum circuits and why this is a need for the method needed to prove quantum advantage (Domingos, 2012).

No practical applications have been shown to have a quantum advantage; this is because current quantum computers are too slow, inefficient, and underpowered to tackle complex real-world issues. Sadly, most claims concerning quantum advantage put out in studies are either based on random circuit sampling or Gaussian boson sampling, neither of which are deemed to be effective applications (Schuld et al., 2014).

The level of background noise in a quantum computer has a direct bearing on how large a problem can be solved Figure 1. This basic chain of events begins with noise, which leads to errors, which prevents you from using as many qubits in your circuit, which in turn restricts the complexity of your algorithm. Obviously, preventing mistakes is crucial (Bonner & Freivalds, 2003).

Error correction studies at IBM date all the way back to 1996, when David DiVincenzo, Charlie Bennett, and John Smolin conducted their own. IBM created the first system in 2015 that could identify quantum bit flip and phase flip mistakes. Almost all commercial and academic quantum computing projects now have some kind of error correcting study underway (Ventura & Martinez, 2000).

Figure 1. Timeline series of quantum computing

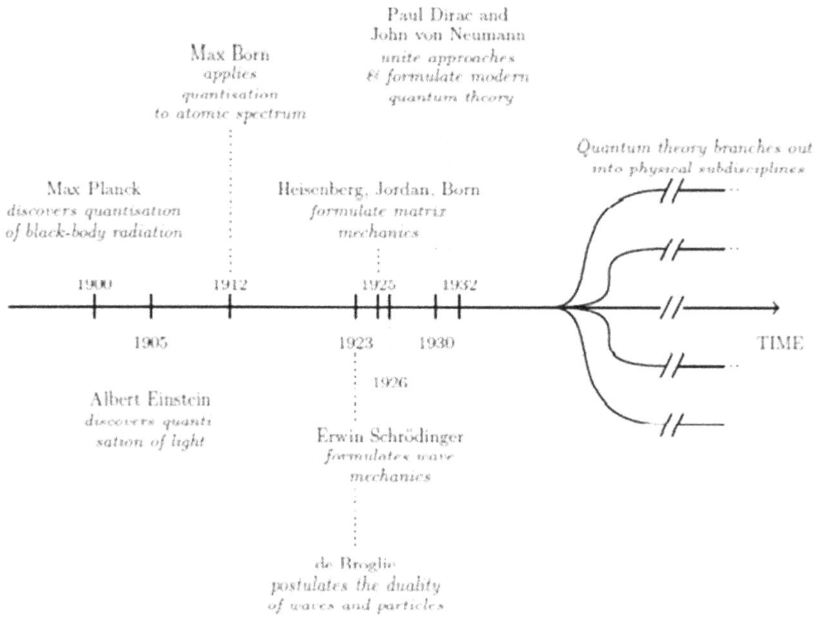

However, quantum error correction (QEC) is a difficult engineering and physics problem, and a workable solution still seems light years away. This is significant because error correction is essential for scaling quantum computers to thousands of qubits (Neven et al., 2009).

However, IBM thinks it is possible to get quantum advantage without fault tolerance. It recently published a study showing that the applications of quantum qubits and gates go far further and deeper than we had anticipated.

The above IBM graph examines the difference between traditional computer scaling and scaling with error-mitigation built into quantum circuits Figure 2. IBM has proven for the first time that quantum error mitigation allows for the efficient execution of increasingly complicated circuits (with only a small increase in simulation cost). While fault-tolerant error correction is still a ways off, it appears that significant speedups are achievable through the employment of error mitigation and error suppression techniques in the meanwhile. In a new publication titled “Evidence for the utility of quantum computing before fault tolerance,” published in Nature, IBM provides proof of this idea. IBM worked with academics from the to generate this report (Lloyd et al., 2013).

Figure 2. Quantum circuit ComplexityIBM

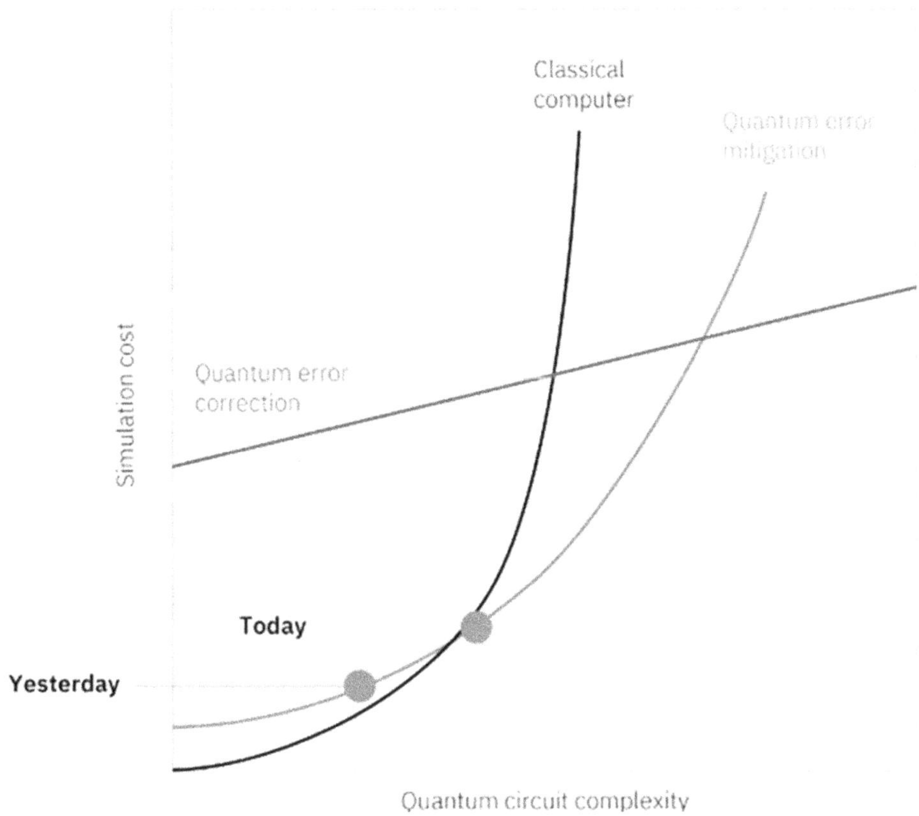

IBM simulated 127 interacting spin states using its 127-qubit Eagle R3 processor. Each qubit in the simulation represented a spin using depth-60 two-qubit gates. With state-of-the-art tensor network methods and high-powered classical supercomputers at the National Energy Research Scientific Computing Centre and Purdue University, UCBerkeley solved a similar problem. The study group purposefully chose a difficult problem that traditional computers struggle with. The study was designed to contrast the quantum answer with the classical one. This was achieved by switching between Clifford and non-Clifford gates and altering the importance of observables in solution attempts between the two groups (Wittek, 2014). The classical supercomputer had trouble with certain parts of the issue where brute force computation was required. In contrast to the supercomputer, the quantum processor kept on solving problems even after it crashed. Since the workflow required a comparison after each iteration, even when a true classical solution was unavailable, sophisticated classical approximation techniques were used to give benchmarks against the quantum outcomes (Schuld et al., 2015). These approximations demonstrated the superior accuracy of the findings produced by the quantum processor compared to the classical approaches. Quantum simulation was still able to operate and give decent results when compared to classical approximation approaches, as shown by its performance against brute-force methods. Quantum computing has enabled more precise results in less time (Bergholm et al., 2018).

Estimates put the power usage for quantum computing at 1.4 MWh per day, while classical supercomputing is estimated to need between 10 and 15 MWh per day. It would appear that quantum also offers a substantial benefit in this area Figure 3.

Figure 3. Timeline feature of quantum with machine learning

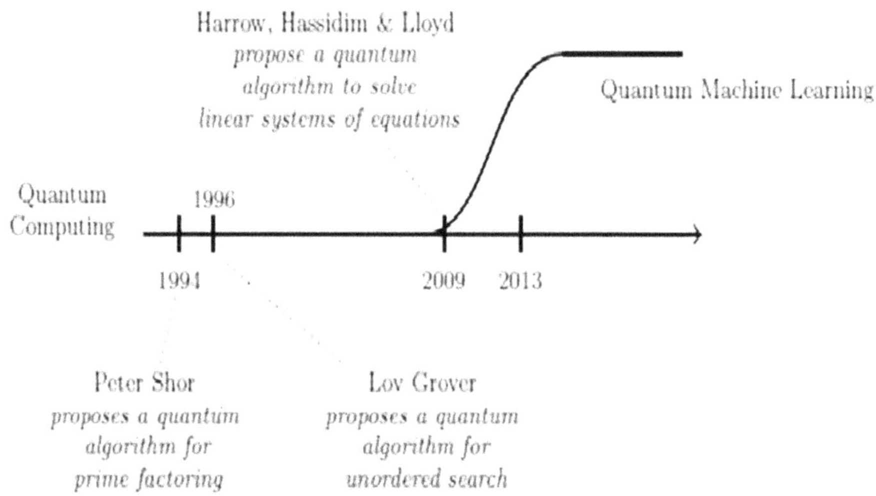

Conventional wisdom holds that fault tolerance is necessary for any quantum computing to be of practical value. Even though IBM's article doesn't prove everything, it does offer a crucial piece of evidence showing that today's quantum computers may deliver value considerably sooner than projected with the help of error mitigation. According to IBM's analysis, the key to achieving a quantum advantage requires a system with at least 100 qubits and 100 gates in depth, as depicted in the above diagram. According to IBM, conclusive proof of quantum advantage may become apparent if sufficient independent study is undertaken in that area of the graph. New error mitigation techniques, such as PEC and ZNE, and high-quality quantum gear from IBM are credited with making this experiment possible. Error mitigation is a key step on IBM's journey towards quantum error correction(Broughton et al., 2020). Once QEC is achieved, we will be able to construct fault-tolerant, multi-million qubit quantum machines optimised for quantum computing. These machines will be able to model and respond to complex financial market behaviour, simulate enormous many-body systems, optimise supply chain logistics, produce new medications and materials, and much more. With the advent of fault-tolerant quantum computers, a new era of research with a quantum focus can begin. With this enhanced power comes the opportunity to effect positive change in the world (Luo et al., 2019).

2. ANALYST NOTES WITH QUANTUM TESTING

IBM sees error mitigation as the most promising approach for closing the gap between the imperfect hardware of today and the fault-tolerant quantum computers of tomorrow. The short-term goal of error mitigation is to facilitate the early attainment of quantum advantage. When compared to other organisations, IBM has conducted the most research on continuously reducing errors (Gilles Brassard & Gambs, 2006).

IBM has already started implementing error mitigation, but the company plans to put more resources into it starting in 2024 so that it may eventually achieve fault tolerance.

Researchers played ping pong with one another during the project we just discussed. Each institution took turns conducting increasingly difficult computations, with some of the work being done at IBM Quantum and some at UC Berkeley. Before running the comparable calculations on the Lawrence Berkeley National Lab and Purdue University supercomputers, IBM Quantum would test the algorithms on the 127-qubit IBM Quantum Eagle processor (Stoudenmire & Schwab, 2016).

However, IBM does not claim that any particular calculation tested on the Eagle processor was faster than what could be accomplished by conventional computers. The IBM-tested computation may soon receive correct responses from other specialised classical methods (Glasser et al., 2018).

2.1 The Top Five Quantum Computers

There are a few distinct ways to design and produce quantum computers right now. The aforementioned list of corporations is by no means complete. The current top five qubit-type methods are:

2.1.1 Superconducting

The superconducting qubit quantum computer is a well-liked variant of the quantum computer. These quantum computers use microscopic electrical circuits to generate and control qubits, and are often constructed from superconducting materials. Superconducting qubits allow for fast gate operations. Google, IBM, IQM, and Rigetti Computing are just a handful of the many companies working on developing and producing superconducting quantum computers (Tang, 2019).

2.1.2 Photonic

In this exclusive interview with The Quantum Insider, Richard Murray, CEO of photonic-based ORCA Computing, describes the nuances and complexity of how these types of quantum computers employ photons (particles of light) to carry and process quantum information. Photonic qubits offer a possible alternative to trapped ions and neutral atoms, which necessitate cryogenic or laser cooling, for use in large-scale quantum computers (Miguel Arrazola et al., 2019).

Dozens of businesses are developing photonic quantum computing systems. Xanadu, ORCA Computing, Quantum Computing, Inc., and PsiQuantum are just a few examples.

An excellent illustration of how quantum computers fall under the umbrella term “Photonics” is the field of “Photonics.” Xanadu’s “squeezed light” technique, for instance, differs significantly from PsiQuantum’s (Carleo & Troyer, 2017).

2.1.3 Elemental Nucleons

Some neutral atom quantum computing providers utilise optical tweezers to hold atoms in place in an ultrahigh vacuum for processing, although this is by no means universal. Because they are less affected by outside electric fields, quantum computers based on neutral atoms are a promising technology (Deng et al., 2017).

Pasqal (merged with Qu&Co), Atom Computing, ColdQuanta, and QuEra are only few of the businesses developing neutral (cold) atom quantum computing technologies.

2.1.4 Ion Traps

In order to store and process quantum information, a trapped ion quantum computer uses atoms or molecules with a net electrical charge, known as "ions," which are trapped and manipulated by electric and magnetic fields. Since trapped ions are shielded from their surroundings, they can be put to use in applications that need extreme steadiness and control. The qubits can also spend a considerable amount of time in a superposition state before losing their coherence.

Quantinuum (a firm formed from the merging of Cambridge Quantum Computing and Honeywell Quantum Solutions), IonQ, Quantum Factory, Alpine Quantum Technologies, eleQtron, and many others make up the trapped ions community of companies in the quantum sector.

Silicon qubits used in a quantum dot computer come in the form of pairs of quantum dots. Such 'connected' quantum dots, in theory, might serve as reliable quantum bits, or qubits, in quantum computers (Giacomo et al., 2018).

Third, The Next Step Beyond Traditional Supercomputing.

Quantum computing has the potential to simulate material components that traditional computers have been unable to model effectively. To solve problems like generating more effective fertilisers, batteries, and medications, the capacity to model these systems is essential. However, current quantum systems are noisy and error-prone, which severely limits their usefulness. This is because quantum bits or qubits are so delicate that they are easily disrupted by their surroundings.

In the latest study, IBM researchers took on a fresh challenge, one that piques the attention of the scientific community's physicists. They employed a 127-qubit quantum processor to model the interaction of 127 atom-scale bar magnets in a magnetic field, a size at which quantum mechanical effects become apparent. The Ising model is a straightforward framework for investigating magnetism (Carrasquilla & Melko, 2017).

Even the most powerful supercomputers would struggle to calculate an exact solution to this problem. The calculation was finished in under a thousandth of a second on the quantum computer. Since quantum calculations are inherently flawed due to the interference and error-inducing effects of quantum noise, they can only be done repeatedly a small number of times.

Both the conventional and quantum methods agreed on the simplest situations where the Ising model could be solved exactly. The quantum algorithm proved to be the correct one when applied to more difficult but still solvable examples.

Thus, “there is reason to believe that the quantum result is more accurate,” as stated by Sajant Anand, a graduate student at Berkeley who undertook much of the work on the classical approximations, for other circumstances where the quantum and classical computations disagreed and no exact answers are known (Ambainis, 2000).

For the Ising model, it is not completely obvious that quantum computing is superior than classical methods.

Error correction is a novel strategy to spotting and fixing mathematical errors; if it proves successful, it will pave the way for widespread adoption of quantum computers. Traditional computers and methods of transmitting data already include error correction to correct for mistakes. However, stronger processors capable of processing many more qubits will be needed before quantum computers can include error correction (Biham et al., 1999).

3. FOR EXPERIMENTAL PHYSICISTS, QUANTUM FRUSTRATION LEADS TO FUNDAMENTAL DISCOVERY

Particles in any system typically collide with one another, resulting in predictable effects similar to the predictable behaviour of billiard balls when they collide. There is a link between the effects and the particles, to put it another way. Some of the limitless possibilities resulting from particle interactions in a frustrated quantum system can give rise to novel quantum states, such as the billiard ball levitating or speeding off at an inconceivable angle Figure 4.

If the number of electrons in the top layer and holes in the bottom layer were equal, then you would expect to see the particles acting in a correlated manner, but Sedrakyan and his colleagues designed the bottom layer so that there is a local imbalance between the number of electrons and holes in the bottom layer. “It’s like a game of musical chairs,” Sedrakyan says, “designed to frustrate the electrons. Instead of each electron having one chair to go to, they must now scramble and have many possibilities in where they ‘sit.’ This frustration kicks off the novel chiral edge state, which has a number of surprising characteristics (Brassard et al., 2000). For instance, if you cool quantum matter in a chiral state down to absolute zero, the electrons freeze into a predictable pattern, and the emergent charge-neutral particles in this state will all either spin clockwise or counterclockwise. Even if you smash another particle into one of these electrons, or you introduce a magnetic field, you can't alter its spin—it's surprisingly robust and can even be used to encode digital data in a fault-tolerant way.

Even more surprisingly is what happens when an outside particle does smash into one of the particles in the chiral edge state. To return to the billiard-ball metaphor, you would expect to send the eight-ball flying when the cue ball smacks into it. But if the pool balls were in a chiral Bose-liquid state, all 15 of them would react in exactly the same way when the eight-ball was struck. This effect is due to the long-range entanglement present in this quantum system (Watrous, 2018).

Figure 4. Frustrated quantum system like rotation chair

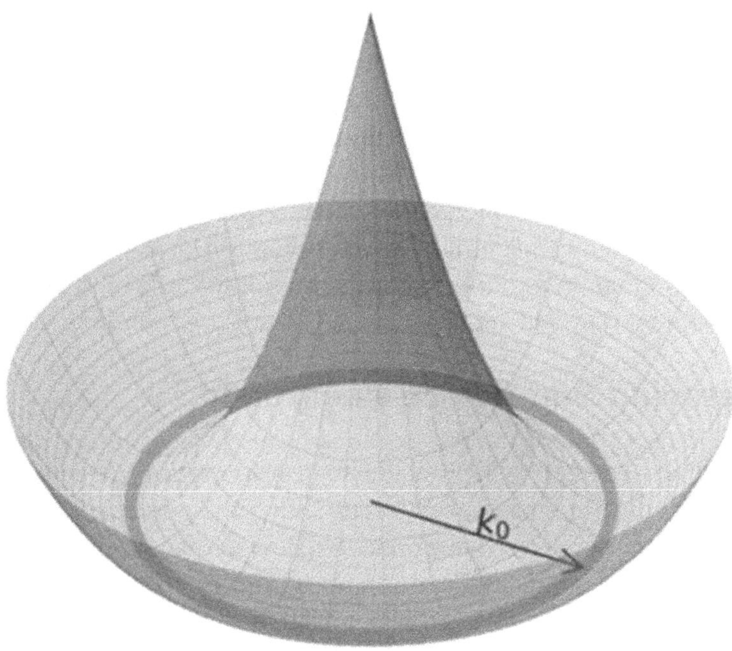

It is difficult to observe the chiral Bose-liquid state, which is why it has remained hidden for so long. To do so, the team of scientists, including theoretical physicists Rui Wang and Baigeng Wang (both of Nanjing University) as well as experimental physicists Lingjie Du (Nanjing University) and Rui-Rui Du (Peking University) designed a theory and an experiment that used an extremely strong magnetic field that is capable of measuring the movements of the electrons as they race for chairs (Abrams & Lloyd, 1999). "On the edge of the semiconductor bilayer, electrons and holes move with the same velocities," says Lingjie Du. "This leads to helical-like transport, which can be further modulated by external magnetic fields as the electron and hole channels are gradually separated under higher fields."

4. DEVICE LEVEL QUANTUM-COMPUTING-HARDENED ENCRYPTION KEYS

In order to provide quantum computing hardened cyber protection for a broad spectrum of connected devices, Quantinuum has announced the debut of Quantum Origin Onboard, an innovation in cryptographic key generation that optimises the strength of keys generated within the devices themselves. Organisations are more vulnerable to cyber threats. Even the cryptographic underpinnings of cybersecurity safeguards remain vulnerable to emerging threats as fraudsters identify new strategies to exploit linked systems and their data. The present standard procedures used by organisations around the world to generate cryptographic keys are not provably unpredictable, putting at risk the security of encrypted data and systems. There is less of a chance that firms will develop and employ weak encryption keys to protect sensitive data thanks to Quantinuum's quantum-computing-hardened cryptographic key improvement (David Clader et al., 2013).

4.1 Improving End-Device Encryption

From essential services to healthcare to transportation to energy, linked devices are driving progress in every sector. Since their prevalence has grown, cyberattacks on networks of interconnected devices have caused widespread damage. The Mirai malware, which spread through infected IP cameras and simple home routers to form a botnet that eventually compromised a major Domain Name System provider, is often cited as an illustration of the destructive potential of IoT-based assaults (Peruzzo et al., 2014). In order to protect connected devices from sophisticated cyberattacks, Quantum Origin Onboard incorporates a cutting-edge security capability inside them. It is the only software solution of its kind, therefore it streamlines deployment and guarantees continuous key generation for all managed information transmitted through networked devices(Farhi et al., 2014).

5. DATA CENTER DECOMMISSIONING TRENDS

It is becoming increasingly important to decommission ageing data centre hardware as data centres expand and upgrade their systems. Consistent with this pattern, quantum technology is driving the use of environmentally friendly decommissioning methods. The company is committed to reducing its environmental effect and protecting the confidentiality of its customers' personal information by emphasising the recycling and proper disposal of electronic waste.

Conserving Energy: Quantum technology places a premium on energy-efficient decommissioning solutions because data centres consume such a large percentage of global energy. During decommissioning, the firm makes use of cutting-edge power management strategies like virtualization and intelligent automation to cut down on energy consumption and carbon emissions as much as possible.With the rise of cloud and edge computing, traditional data centres are becoming obsolete and must be shut down. Using quantum technology, businesses can safely and efficiently move their infrastructure into the cloud, improving their resource utilisation and cutting down on operational costs in the process(Farhi & Harrow, 2016).

When compared to what is required for practical applications, "all the quantum computing machines out there right now are very basic," Pribiag added. "Scaling up is required if we are to build a computer capable of solving practical, high-level issues. Many scientists are currently investigating methods and potential applications for computers or AI devices that could one day surpass the performance of classical computers. We're working on hardware for quantum computers that would make these algorithms practical.

A diode is an electrical component that permits current to flow in one direction in a circuit but not the other. It's basically the opposite terminal of a transistor, the primary component of modern computer chips. Researchers are interested in developing diodes out of superconductors because of its ability to transport energy without any power being lost (Farhi et al., 2000; Verdon et al., 2017).

Unlike regular diodes, which can only process one input and one output, their gadget can process many signals simultaneously. This capability may find use in neuromorphic computing, a technique for improving the efficiency of artificial intelligence systems that involves designing electrical circuits to replicate the way neurons behave in the brain.

The researchers' approach is more generalizable than other methods in the field because it can, in theory, be utilised with any kind of superconductor.

These features make their device more applicable to commercial settings and hold promise for accelerating the widespread adoption of quantum computers.

REFERENCES

Abrams, D. S., & Lloyd, S. (1999). Quantum algorithm providing exponential speed increase for finding eigenvalues and eigenvectors. *Physical Review Letters*, *83*(24), 5162–5165. doi:10.1103/PhysRevLett.83.5162

Ambainis, A. (2000). Quantum lower bounds by quantum arguments. In *Proceedings of the Thirty-Second Annual ACM Symposium on Theory of Computing,* (pp. 636–643). ACM. 10.1145/335305.335394

Bergholm, V., Izaac, J., Schuld, M., Gogolin, C., Blank, C., McKiernan, K., & Killoran, N. (2018). *Pennylane: Automatic differentiation of hybrid quantum-classical computations*. arXiv preprint arXiv:1811.04968

Biham, E., Biham, O., Biron, D., Grassl, M., & Lidar, D. A. (1999). Grover's quantum search algorithm for an arbitrary initial amplitude distribution. *Physical Review A*, *60*(4), 2742–2745. doi:10.1103/PhysRevA.60.2742

Bonner, R., & Freivalds, R. (2003). A survey of quantum learning. Quantum Computation and Learning, 106.

Brassard, G., Høyer, P., Mosca, M., & Tapp, A. (2000). Quantum amplitude amplification and estimation. *Contemporary Mathematics*, *305*, 53–74. doi:10.1090/conm/305/05215

Broughton, M., Verdon, G., McCourt, T., Martinez, A. J., Yoo, J. H., Isakov, S. V., Massey, P., & Niu, Y. (2020). *Tensorflow quantum: a software framework for quantum machine learning*. arXiv preprint arXiv:2003.02989.

Carleo, G., & Troyer, M. (2017). Solving the quantum many-body problem with artificial neural networks. *Science*, *355*(6325), 602–606. doi:10.1126cience.aag2302 PMID:28183973

Carrasquilla, J., & Melko, R. G. (2017). Machine learning phases of matter. *Nature Physics*, *13*(5), 431–434. doi:10.1038/nphys4035

David Clader, B., Jacobs, B. C., & Sprouse, C. R. (2013). Preconditioned quantum linear system algorithm. *Physical Review Letters*, *110*(25), 250504. doi:10.1103/PhysRevLett.110.250504 PMID:23829722

Deng, D.-L., Li, X., & Das Sarma, S. (2017). Quantum entanglement in neural network states. *Physical Review X*, *7*(2), 021021. doi:10.1103/PhysRevX.7.021021

Domingos, P. (2012). Afewuseful things to knowabout machine learning. *Communications of the ACM*, *55*(10), 78–87. doi:10.1145/2347736.2347755

FarhiE.GoldstoneJ.GutmannS. (2014). A quantum approximate optimization algorithm. arXiv:1411.4028

Farhi, E., Goldstone, J., Gutmann, S., & Sipser, M. (2000). Quantum computation by adiabatic evolution. arXiv:quant-ph/0001106. MIT-CTP-2936.

FarhiE.HarrowA. W. (2016). Quantum supremacy through the quantum approximate optimization algorithm. arXiv:1602.07674

Giacomo, T., Guglielmo, M., Juan, C., Matthias, T., Roger, M., & Giuseppe, C. (2018). Neural-network quantum state tomography. *Nature Physics*, *14*(5), 447–450. doi:10.103841567-018-0048-5

Gilles Brassard, E. A., & Gambs, S. (2006). Machine learning in a quantum world. In *Advances in Artificial Intelligence* (pp. 431–442). Springer.

Glasser, I., Pancotti, N., & Cirac, J. N. (2018). Supervised learning with generalized tensor networks. arXiv preprint arXiv:1806.05964.

Lloyd, S., Mohseni, M., & Rebentrost, P. (2013). Quantum algorithms for supervised and unsupervised machine learning. arXiv preprint arXiv:1307.0411

Luo, X.-Z., Liu, J.-G., Zhang, P., Wang, L., & Yao, J. (2019). Extensible, efficient framework for quantum algorithm design. arXiv preprint arXiv:1912.10877.

Miguel Arrazola, J., Delgado, A., Bardhan, B. R., & Lloyd, S. (2019). *Quantum-inspired algorithms in practice*. arXiv preprint arXiv:1905.10415.

Neven, H., Denchev, V. S., Rose, G., & Macready, W. G. (2009). Training a large scale classifier with the quantum adiabatic algorithm. arXiv preprint arXiv:0912.0779.

Nielsen, M. A., & Chuang, I. L. (2010). *Quantum Computation and Quantum Information*. Cambridge University Press.

Peruzzo, A., McClean, J., Shadbolt, P., Yung, M.-H., Zhou, X.-Q., Love, P. J., Aspuru-Guzik, A., & Obrien, J. L. (2014). A variational eigenvalue solver on a photonic quantum processor. *Nature Communications*, *5*(1), 5. doi:10.1038/ncomms5213 PMID:25055053

Schuld, M., Sinayskiy, I., & Petruccione, F. (2014). The quest for a quantum neural network. *Quantum Information Processing*, *13*(11), 2567–2586. doi:10.100711128-014-0809-8

Schuld, M., Sinayskiy, I., & Petruccione, F. (2015). Introduction to quantum machine learning. *Contemporary Physics*, *56*(2), 172–185. doi:10.1080/00107514.2014.964942

Stoudenmire, E., & Schwab, D. J. (2016). Supervised learning with tensor networks. Advances in Neural Information Processing Systems, 4799–4807.

Tang, E. (2019). Aquantum-inspired classical algorithm for recommendation systems. In *Proceedings of the 51st Annual ACM SIGACT Symposium on Theory of Computing* (pp. 217–228). ACM. 10.1145/3313276.3316310

Ventura, D., & Martinez, T. (2000). Quantum associative memory. *Information Sciences*, *124*(1), 273–296. doi:10.1016/S0020-0255(99)00101-2

VerdonG.BroughtonM.BiamonteJ. (2017). A quantum algorithm to train neural networks using low-depth circuits. arXiv:1712.05304

Watrous, J. (2018). *Theory of Quantum Information*. Cambridge University Press. doi:10.1017/9781316848142

Wittek, P. (2014). *Quantum Machine Learning: What Quantum Computing Means to Data Mining*. Academic Press.

Chapter 3
Spintronics System:
Spin Polarization and Optical Characterization

Shivakumar Hunagund
https://orcid.org/0000-0001-5201-5827
Intel Corporation, USA

ABSTRACT

Spintronic systems, which involve manipulating the spin of electrons, have potential for use in quantum computing due to their unique properties. These systems can create spin-based qubits that are more stable and less prone to decoherence than other types, and can also be used to control interactions between qubits. They can also utilize spin-based quantum gates that can perform specific operations faster and more efficiently. Spintronic systems hold promise for the development of powerful and efficient quantum computers.

INTRODUCTION

Spintronic systems have the potential to be used for quantum computing due to the unique properties of electron spin. Quantum computers rely on the ability to control and manipulate quantum states, and the spin of an electron can serve as a qubit, or a unit of quantum information. By manipulating the spin of an electron, it is possible to perform quantum operations and perform calculations that would be otherwise impossible on a classical computer.

DOI: 10.4018/978-1-6684-7535-5.ch003

Copyright © 2023, IGI Global. Copying or distributing in print or electronic forms without written permission of IGI Global is prohibited.

One way in which spintronic systems can be used for quantum computing is using spin-based qubits. These qubits are formed by manipulating the spin state of an electron, and they have the potential to be more stable and less susceptible to decoherence than other types of qubits. In addition, spintronic systems can be used to manipulate and control the interactions between qubits, which is an essential component of quantum computing.

Another potential application of spintronic systems in quantum computing is through the use of spin-based quantum gates. These gates operate on the spin states of electrons to perform specific quantum operations, and they have the potential to be faster and more efficient than other types of quantum gates.

Overall, the use of spintronic systems for quantum computing is an active area of research, and it holds great promise for the development of powerful and efficient quantum computers.

In this chapter, we will explore the theoretical and experimental aspects of the spin manipulation and optical read out in a Spintronic system. Spintronics (spin electronics), is a study of the electron's spin degree of freedom and it's associated magnetic moment, in addition to its fundamental electronic charge, in solid state physics. The use of Semiconductors spintronics originated in the late 80s by Datta and Das in their theoretical proposal of a spin field-effect-transistor. In their proposed electro optic light modulator, the current modulation in the suggested structure arises from spin precession due to the spin orbit coupling in narrow gap semiconductors, while magnetized contacts are used to preferentially inject and detect specific spin orientations. While spintronics has numerous potential applications, including data storage, its most promising use is in the realm of quantum information processing. This chapter aims to provide a comprehensive overview of these concepts and their relevance in the field of spintronics.

All spintronic devices act according to the simple scheme:

1. Information is stored (written) into spins as a particular spin orientation (or ⁻),
2. The spins, being attached to mobile electrons, carry the information along a material, and
3. The information is read at a terminal.

Spin orientation of conduction electrons survives for a relatively long time (nanoseconds) i.e. decoherence is relatively slow, compared to tens of femtoseconds during which electron momentum decays, which makes spintronic devices particularly attractive for quantum information processing and quantum computation where electron spin would represent a bit (called qubit) of information.

The electron spin (or ¯) state can be used to represent a classical bit with a logical (1 or 0) and any quantum superposition of these. The general state is expressed as,

$$|\Psi> = a|0> + b|1>$$

Figure 1. The Bloch sphere is a representation of a qubit, the fundamental building block of quantum computers

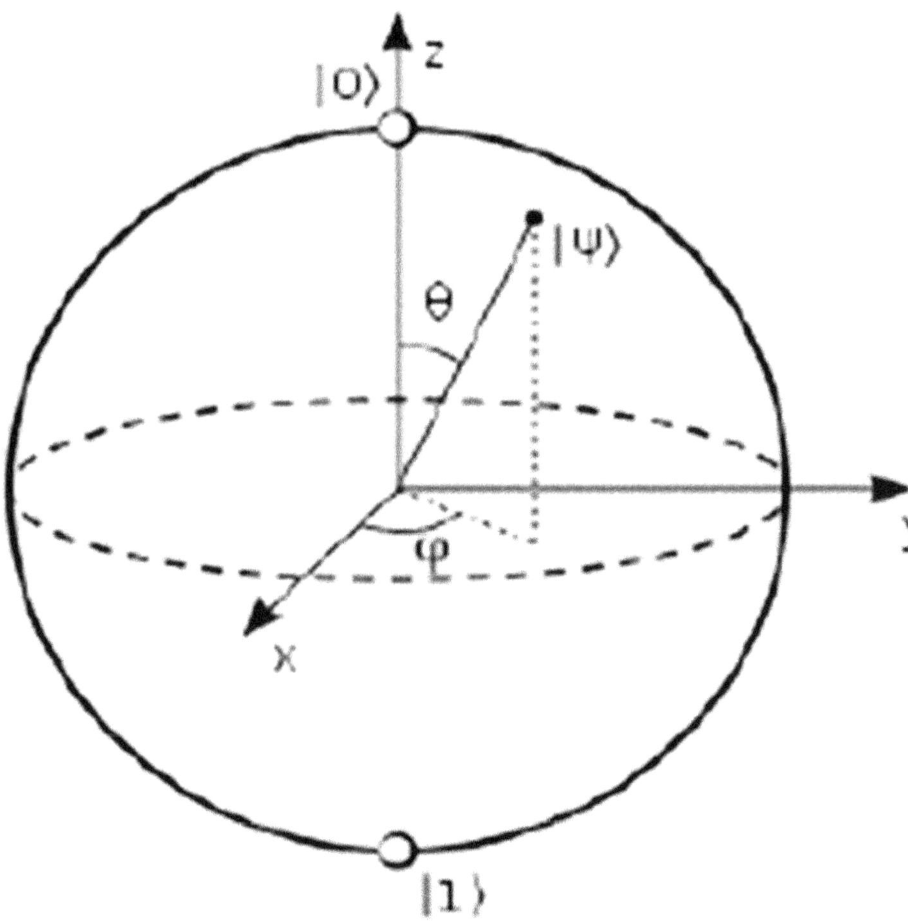

i.e., as a superposition of both states. Thus a measurement of the qubit will cause it's wavefunction to collapse into the state $|0>$ with probability $|a|^2$ or into the state $|1>$ with probability $|b|^2$. This means that during its time evolution a qubit may be partly in both the $|0>$ and $|1>$ state at the same time, i.e., to the degree that a and b may adopt an infinity of values, the qubit has the potential to be in any of these.

A spin based quantum information process needs the following pre-requisites-

- Initialization of the spin states
- Storage at well-defined sites
- Techniques to manipulate spins and finally
- Read out the result of the performed calculations

In contrast to classical computation where the information unit, the bit, can be read and copied at anytime, quantum mechanics forbids such things: There is no-cloning theorem and spying at the qubit destroys its coherences. A single microwave photon can destroy the coherences of a Rb atom passing a double slit (Loeffler, 2008). Therefore the system needs to be well isolated from the environment. InGaAs/GaAs quantum dots has been identified as a promising candidate for quantum information storage due to their long spin coherence times for electrons and excitons. A spin-polarized state can be provided and optical readout of spin states is possible by observing the recombination radiation. In this experiment, ZnMnSe - a diluted magnetic semiconductor is used as a spin aligner. Quantum information processing requires high initialization fidelities and the ability to address single spin-qubits stored at individual localized sites. A concurrent electrical initialization of several spin-qubits is accomplished with polarization degrees close to 100% by electrical spin injection from the diluted magnetic semiconductor ZnMnSe into InAs/GaAs quantum dots (Hettrich, n.d.; Loeffler, 2008). Individual spin states in single quantum dots can be optically addressed and read out through metallic (gold) nano-apertures.

Figure 2. Spin-LED structure

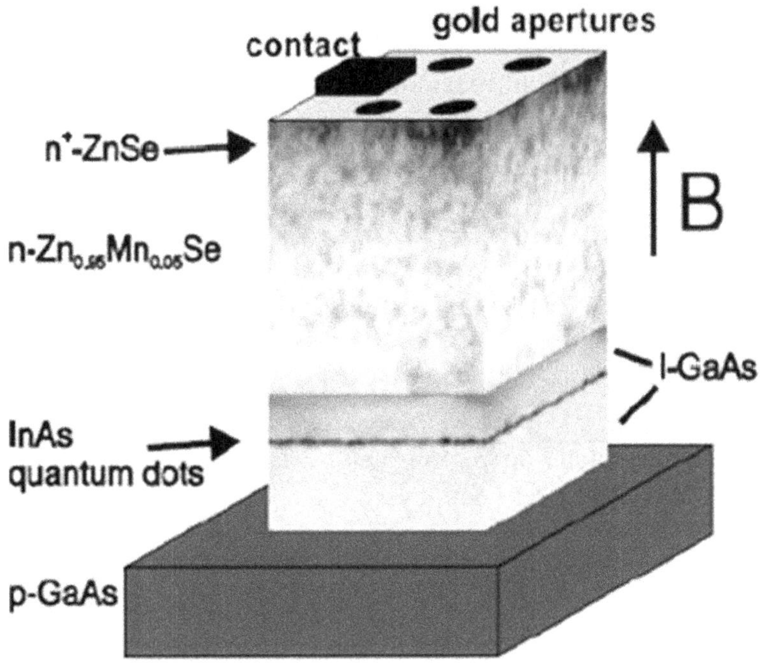

PREPARATION AND SETUP

It is advantageous to use semiconductor materials not only because of well-controlled fabrication techniques but the special band structure enables coupling to the light field via interband transitions. Thus spin information can be coherently transportable and might allow coupling of distant qubits. Moreover, the experiment involves all-electrical preparation of spin states where it is easy to combine with the current semiconductor technology and also reduce the need for laser systems.

The electron spin is controlled via use of magnetic fields. They first pass the spin aligner material (ZnMnSe) before being injected into the InAs/GaAs dots. These C are grown using molecular beam epitaxy (MBE) machine following these set of conditions (Loeffler, 2008; Merz, 2014):

The growth temperature must be compatible: GaAs grows at above 550° C and ZnMnSe at below 400° C. So, growing ZnMnSe on GaAs is possible but not the other way around. Growing GaAs on ZnSe substrate will lead to desorption at GaAs growth temperature.

The crystal system must match: The purpose is to deposit the spin aligner (ZnMnSe) on GaAs, which crystallizes in the cubic system in Zincblende structure. Crystallization of ZnMnSe depends on manganese concentration.

The lattice constants should be similar: GaAs has an in-plane lattice constant of a_0 = 5.65325 Å and $Zn_{0.95}Mn_{0.05}Se$ with 5% manganese concentration has a_0 = 5.68121 Å. The lattice constant can be calculated with a_0, $Zn_{1-x}Mn_xSe$ = (Ö2)×(4.009+0.1645*x*), where *x* is the manganese concentration.

E.g.: For 0.5% Manganese concentration:

a_0, $Zn_{0.95}Mn_{0.05}Se$ = (Ö2)×(4.009+(0.1645×0.5))

= 5.68121 Å

The lattice mismatch *f* on GaAs is:
f= asubstrate – aepilayer / aepilayer

= (5.68121−5.65325)÷5.65325 = 0.5%.

Depending on the lattice mismatch, the epilayer grows fully strained on the substrate without dislocations or the epilayer relaxes via formation of dislocations to its intrinsic lattice constant. The thickness at which this happens is called critical thickness. The critical thickness for ZnSe on GaAs is about 200nm. At higher Mn concentration a lower external magnetic field is needed to obtain the same spin polarization. However, at 13% Mn concentration the measured spin-polarization of $Zn_{0.87}Mn_{0.135}Se$ is lower than that of 5% Mn concentration $Zn_{0.95}Mn_{0.05}Se$ due to high dislocation density. This problem can be overcome by introducing Sulfur into the material. Sulfur decreases the lattice constant of the quaternary material ZnMnSSe in comparison to ZnMnSe. An ideal composition is found out to be $Zn_{0.87}Mn_{0.13}S_{0.17}Se_{0.83}$.

Now the subsequent step in the development of semiconductor heterostructure is the production of quantum dots that address the requirement of information storage. An efficient way to produce large amounts of quantum dots is via self-assembly. Self-assembled quantum dots nucleate spontaneously under certain conditions during molecular beam epitaxy (MBE), when a material is grown on a substrate to which it is not lattice matched. The resulting strain produces coherently strained islands on top of a two-dimensional wetting layer. This growth mode is known as Stranski–Krastanov growth. The islands can be subsequently buried to form the quantum dot.

Figure 3. (Far-left) CELFA (composition estimation by lattice fringe analysis) TEM image of quantum dots in indium rich islands within a 2D layer of InGaAs quantum as well; (left) AFM image of uncapped quantum dot sample

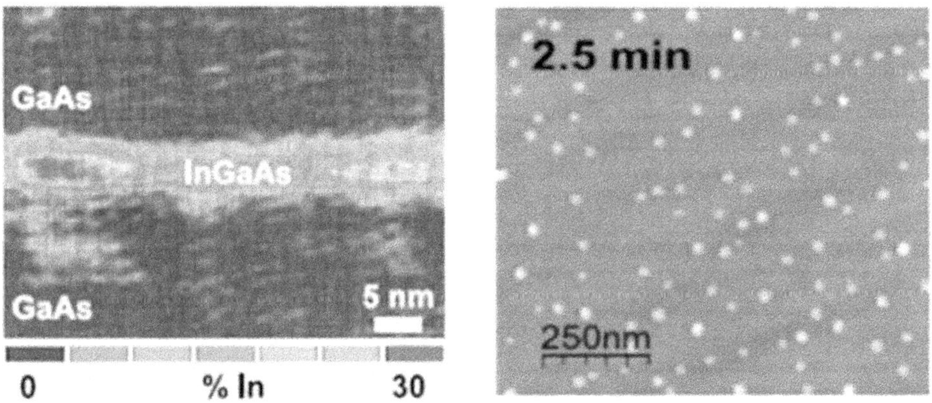

InAs (dot material) deposited on GaAs (matrix material) adopts its lattice constant which leads to a strained layer. Continuous deposition of InAs leads to formation of small islands at critical thickness in order to minimize surface energy which is thermodynamically more favorable. On further deposition, indium atoms diffuse over sample surface and larger islands are formed. The confinement of the electron in all three spatial dimensions is due to the conduction and valence band discontinuity between GaAs and $Ga_{1-x}In_xAs$. Electrons and holes are confined in the indium rich regions.

After the growth of the quantum dots, they are capped with intrinsic GaAs at the InAs growth temperature. The thickness of the cap layer is chosen to be 25nm which is sufficiently large enough to avoid any influence of the magnetic spin aligner layer on electrons in the quantum dots. An indium contact pad is structured by standard lithography process with the n-ZnSe contact layer for electrical conductivity.

A thin gold layer was thermally evaporated wherein apertures were defined by electron beam lithography. Gold nano-apertures are fabricated using ebeam lithography process. Mesa structuring (square-shaped spin-LEDs) is done via photolithography and etched by a two step process – first to remove the II-VI layer using diluted $K_2Cr_2O_7$ combined with Hbr and then remove the III-V layer with higher concentration of $K_2Cr_2O_7$. Now the final step involves packing the sample for experiment. A Silver conducting glue is used to fix the sample on a copper sample holder which ensures good heat and electrical conductivity. A 25μm gold wire is bonded using a conducting epoxy resin with the top indium contacts of the sample.

Figure 4. Formation of quantum dots (of inhomogeneous size) by MBE in Stranski-Krastonow growth mode. By adjusting the MBE growth parameters, the emission energy of the quantum dots can be adjusted over a broad wavelength range. Those parameters include – the amount of InAs deposited, magnitude of the In diffusion on the sample and a possible subsequent annealing of the capped sample. Another method is by changing the confinement by altering the composition of the barriers.

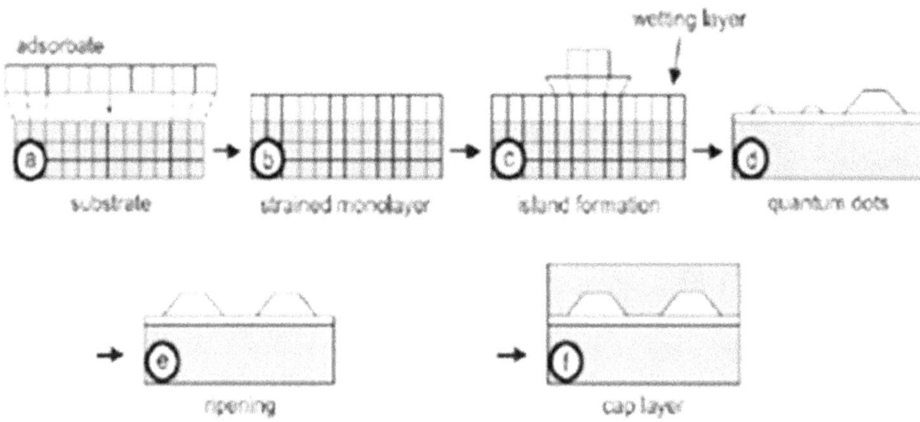

Figure 5. Left to right: Image of the device with the chip in the copper holder; optical microscope image of a single cell mesa with gold apertures; electron microscope image of a single aperture

SPIN POLARIZATION OF ELECTRONS AND INJECTION INTO QUANTUM DOTS

The basic operation principle of a spin-LED is similar to p-i-n diode. An appropriate voltage is applied to the sample and the unpolarized electrons from the top contact get polarized due to a giant Zeeman splitting of the spin up and spin down states in the spin aligner ZnMnSe layer before they are injected into the quantum dots. The holes entering into the quantum dots from the bottom contact do not have a defined spin polarization. Due to electron-hole recombination, circularly polarized photons are emitted. The helicity of the photon is directly related to the spin-polarization of the electron.

Figure 6. (Left) Basic structure of InGaAs quantam-dot spin-LEDs used in the experiment. (Right) Flat band diagram showing spin-LEDs principle of operation.

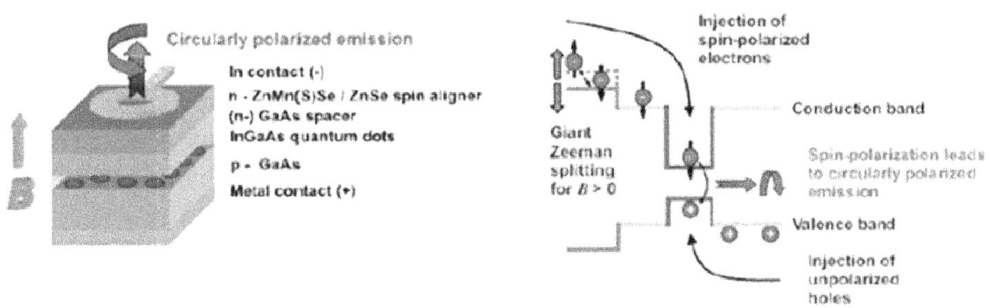

The first step is the polarization of electrons and in here a diluted magnetic semiconductor $Zn_{1-x}Mn_xSe$ is used to generate spin-polarized electrons (spin aligner). 'It renders traversing electrons spin-polarized when a magnetic field is applied. This is due to electrons relaxing into the energetically lower of the two spin-split conduction bands, which are separated through a giant Zeeman splitting. Thus, applying a voltage across the spin-LED results in spin-polarized electrons reaching the QD, where they recombine with unpolarized holes injected from the bottom part of the spin-LED. Due to optical selection rules, these transitions can only take place under the emission of circularly polarized light. A submicron aperture on top of the heterostructure helps to minimize spurious emission from other nearby QDs' (Asshoff et al., 2011).

With electrical spin injection many qubits in different QDs can be initialized simultaneously. This would be difficult to achieve with all-optical techniques (e.g., resonant excitation of the dots with circularly polarized light), because the involved electronic transition energies vary from dot to dot.(Hettrich,) Unpolarized holes are fed into the dots from the bottom p-GaAs layer and duc to thc strong strain induced heavy-hole/light-hole splitting, only the $\pm 3/2$ heavy-hole QD states are populated and lead to optical transitions. Electrons with spin polarization $-1/2$ can only recombine with $-3/2$ and similarly electrons spin polarized $+3/2$ recombine with $+1/2$ holes, emitting circularly polarized $\sigma+$ or $\sigma-$ photons in Faraday geometry. As measure of the photon polarization state, the circular polarization degree is defined -

$$CPD = (I_{\sigma+} - I_{\sigma-})/(I_{\sigma+} + I_{\sigma-})$$

with $I_{\sigma+(-)}$ denoting the intensity of $\sigma_{+(-)}$ -polarized light. *CPD* indicates the type and degree of electron spin polarization in the QDs.

Figure 7. The g factor in the QDs is opposite to that in ZnMn(S)Se meaning the electrons are injected into the higher energy spin state. Spin-down electrons generated by the spin aligner are injected into the upper electronic/excitonic spin state of the dots.

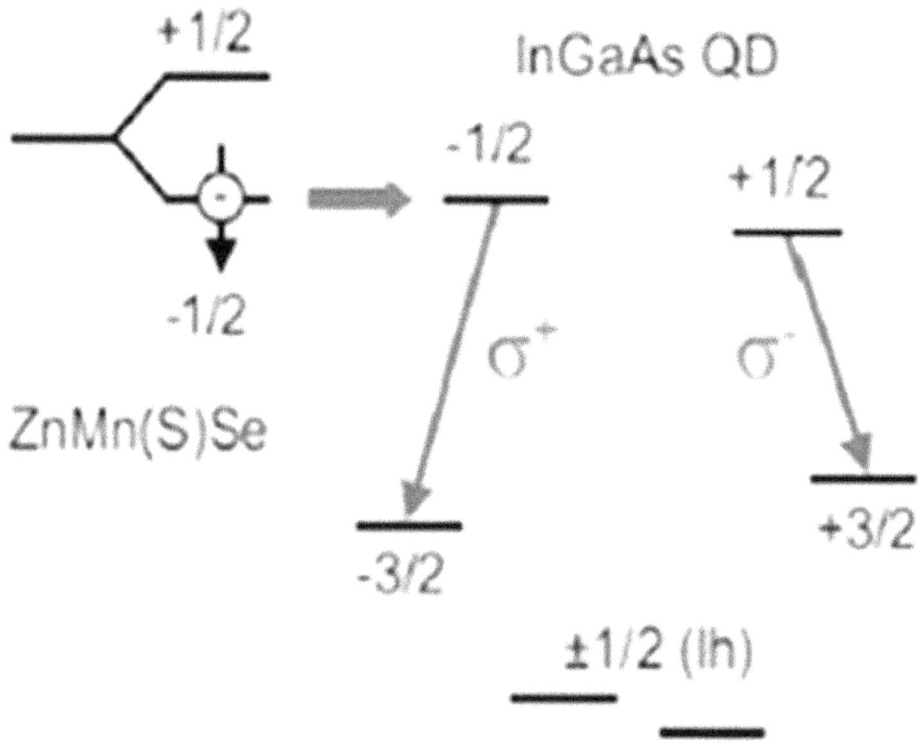

MANIPULATION EXPERIMENT

Experiments were carried out in a magneto-optical cryostat with the sample temperature set to about $T = 5$ K. The sample is placed in a microwave resonator and inserted in a cryostat filled with liquid helium (sample remain isolated from the liquid helium) and measurements were carried out in a superconducting coil magnet cryostat with optical access to the sample established. The applied magnetic field (ranging upto 14T) is perpendicular while the resonating B-field of the microwave pulses are parallel to the plane of the sample. The sample's position can be controlled via a three axis piezo unit so that a single mesa can be positioned in the focus of a 35mm lens which collects the luminescence and guide the light from one aperture of the device outside the cryostat. The polarization selectivity (to differentiate between σ_+- and σ_- -polarized photons) is achieved by passing the luminescence from the sample first through a broadband quarter-waveplate which transforms circular polarization into a linear one, and then selects the desired polarization with a high contrast Glan-laser polarizer. The photons are focused by an aspheric lens on a standard multimode fiber and guided to a spectrometer where a charge-coupled device (CCD) detects the diffracted light.

Figure 8. Micro-electro luminescence (μ-EL) spectra of a single quantum-dot Spin-LED for different applied magnetic fields. I = 4mA with 2.1 volts and 5k temperature. Without the external magnetic field, both components contribute equally to the signal. As the field is increased, the o+ component gets stronger.

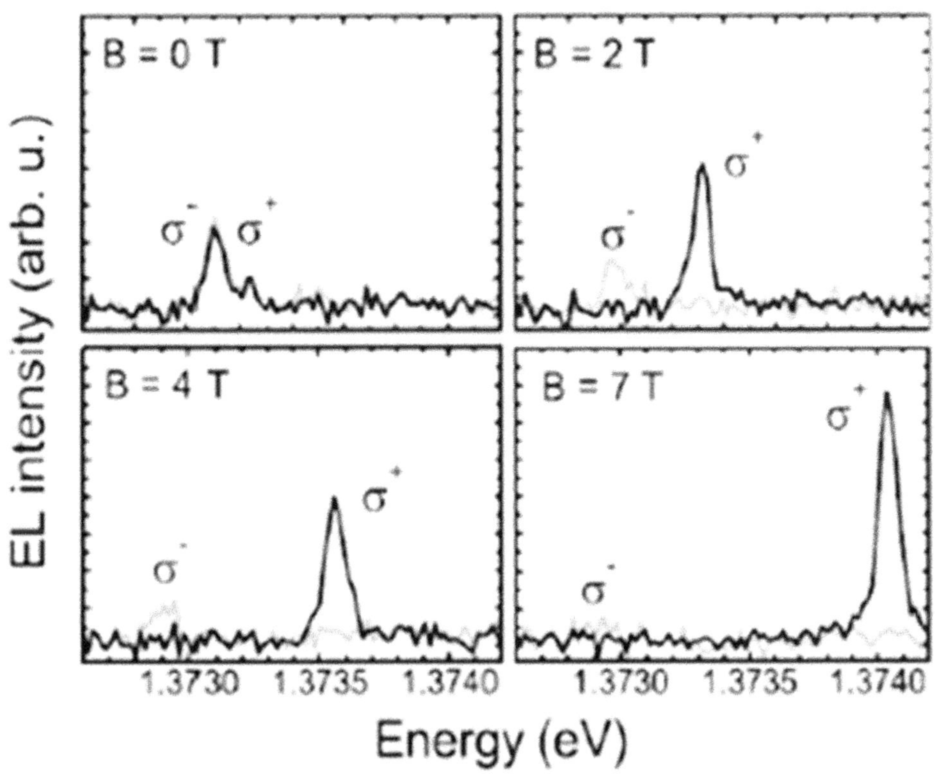

The degeneracy of the spin-up and spin-down sublevels in the QD is lifted when a magnetic field is applied (Zeeman splitting), shifting the σ_+- / σ_- -polarized excitonic emission from the spin-down/spin-up conduction band sublevel to higher/lower energies (Asshoff et al., 2011). At B = 0 T, a sharp emission peak (line width resolution limited) from a single dot with no circular polarization is observed. For non-vanishing magnetic fields, the Zeeman splitting of the QD transition can be observed. When the magnetic field is increased, the electrons injected into the dot become more and more spin-polarized. As a result, the σ+ transition, corresponding to the optically active exciton state for the injected spin-down electrons, grows, while the spin-up-related σ− peak drops strongly. Finally, at about B = 7 T, the σ− emission nearly disappears, indicating that the electrons in the QD are highly spin-polarized.(Hettrich,)

Figure 9. CPD of the EL for different QDs in the same spin-LED. The achieved CPD varies strongly from dot to dot, even within the same device and for similar emission wavelengths, the average polarization increases for higher energy Qds.

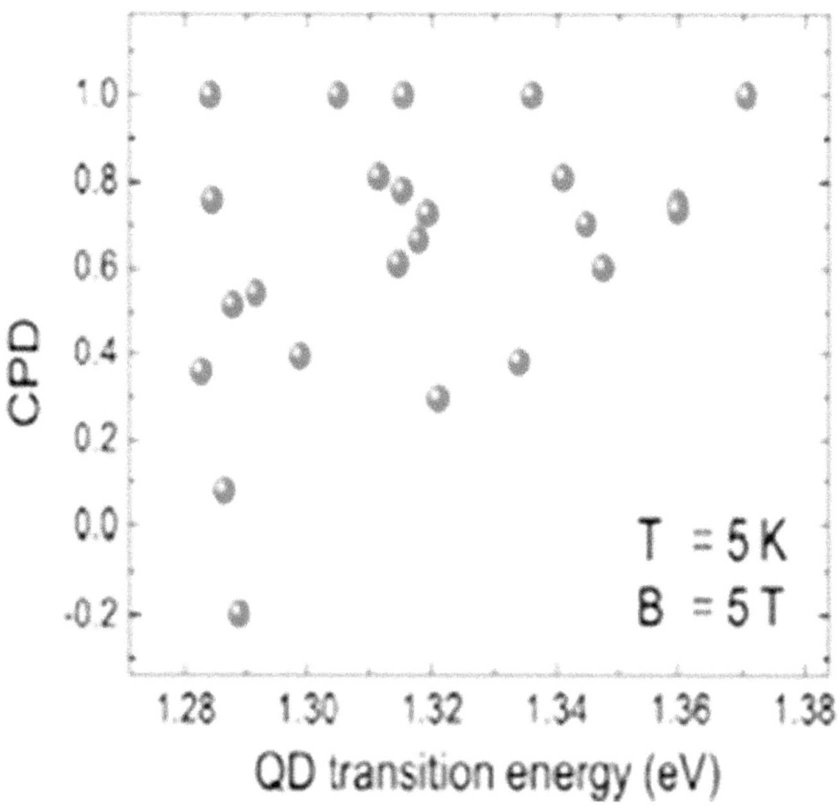

The ability to control the quantum state of a single electron spin in a quantum dot is an important step in achieving a scalable spinbased quantum computer. Commonly used technique for inducing spin flips is electron spin resonance (ESR). ESR is the physical process whereby electron spins are rotated by an oscillating magnetic field B_{ac} (with frequency f_{ac}) that is resonant with the spin precession frequency in an external magnetic field Bext, oriented perpendicularly to B_{ac} ($hf_{ac} = g\mu_B B_{ext}$, with μ_B the Bohr magneton and g the electron spin g-factor). An oscillating magnetic field resonant with the Zeeman splitting can flip the spin in the dot. Spin-manipulation can be enabled via electron spin resonance (ESR) setup. This setup allows for high field ESR (53 GHz) with a tunable high power microwave source. The samples are specially designed in order to fit into a cylindrical H01-resonator to achieve a well-defined microwave field distribution. A pulsed microwaves parallel to the surface plane of the sample is applied and tuned to the best suitable resonant frequency. The pulse are timed – $t_{\pi\text{-pulse}} = \pi / \gamma B_{\perp}$, where $\gamma = g\,\mu B/\hbar$ is the resonant frequency that matches the gap corresponding to the external magnetic field (zeeman splitting).

Figure 10. (Far right) shows the final setup for the experiment before inserting it into the cryostat. (Right) Microwave resonator.

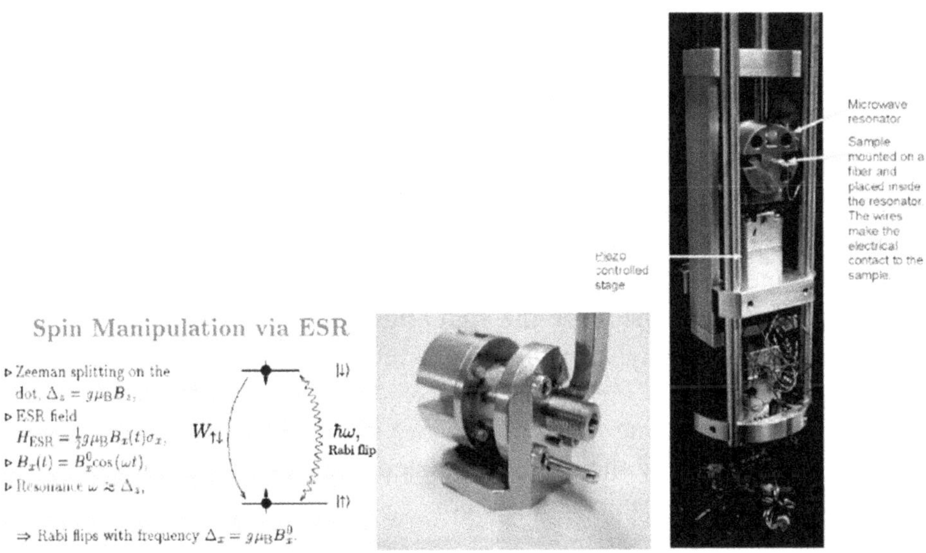

DEVICE OPTIMIZATION

It can be noticed from Figure 9 that the polarization degree for individual quantum dots in the ensemble varies. This can be ascribed to spin loss mechanisms that influence the spin relaxation time. Experimental evidence suggests that the spin polarization is lost before the electrons are captured in the dots. This loss can be attributed to the defect related scattering process which occurs outside of the QDs (quantum dots). Defects due to dislocations at the III-V/II-VI interface can be optimized by reducing the lattice mismatch between $Zn_{1-x}Mn_xSe$ and GaAs. This has been done by developing lattice matched $Zn_{1-x}Mn_xS_ySe_{1-y}$ spin aligners. Figure 11a shows the comparative measurements of two spin LEDs. The use of ZnMnSSe spin aligner approximately lattice-matched to GaAs significantly improves the achieved CPD. Introducing higher Mn concentration in lattice-matched spin aligners would provide larger zeeman splitting at lower magnetic fields and/or higher temperatures. Fig. *9* also indicates that average polarization increases for higher energy QDs. This could be explained in terms of peculiarities in band structure of the device and the phenomena of electron tunneling through a barrier in the quantum dots *(Ref* Figure 11b). At the III-V/II-VI interface, the band bending and the position of the electronic quasi- Fermi level $E_F(e)$ give rise to the formation of a two-dimensional electron gas (2DEG). Also, the injected spin-polarization electrons have to tunnel through a potential barrier in order to reach the wetting layer and the quantum dots. Since the average effective barrier is larger for low-energy dots, tunneling will be slower causing spin relaxation and therefore CPD drops for these quantum dots. Shifting the electron Fermi level to lower energies would prevent formation of 2DEG at the III-V/II-VI interface and the injection of poorly polarized electrons into the QDs. This could be realized by raising the effective QD density of states via an increased dot density. Spin relaxation occurring in the GaAs spacer layer between the spin aligner and the QDs which results in slow electron tunneling into QDs can be reduced by decreasing the spacer thickness (Figure 11c). Furthermore, Zeeman splitting in the spin aligner decreases with increasing temperature *(Ref* Figure 11d) resulting in upper electron spin level being thermally populated, thus leading to a lower initial spin polarization of the injected barriers.

Figure 11. a) CPD measured for different spin-aligner material under magnetic field, b) Spin-LED band structure, c) CPD dependence on spacer thickness, and d) temperature

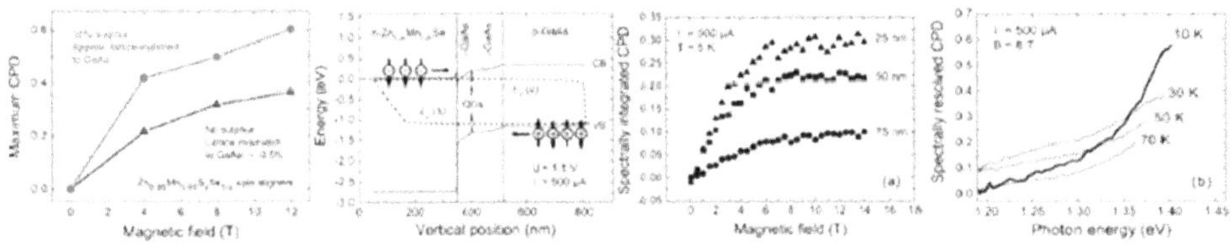

Figure 12. III-V structure with InGaAs quantum dots, II-VI spin aligner $Zn_{0.98}Mn_{0.016}Se$

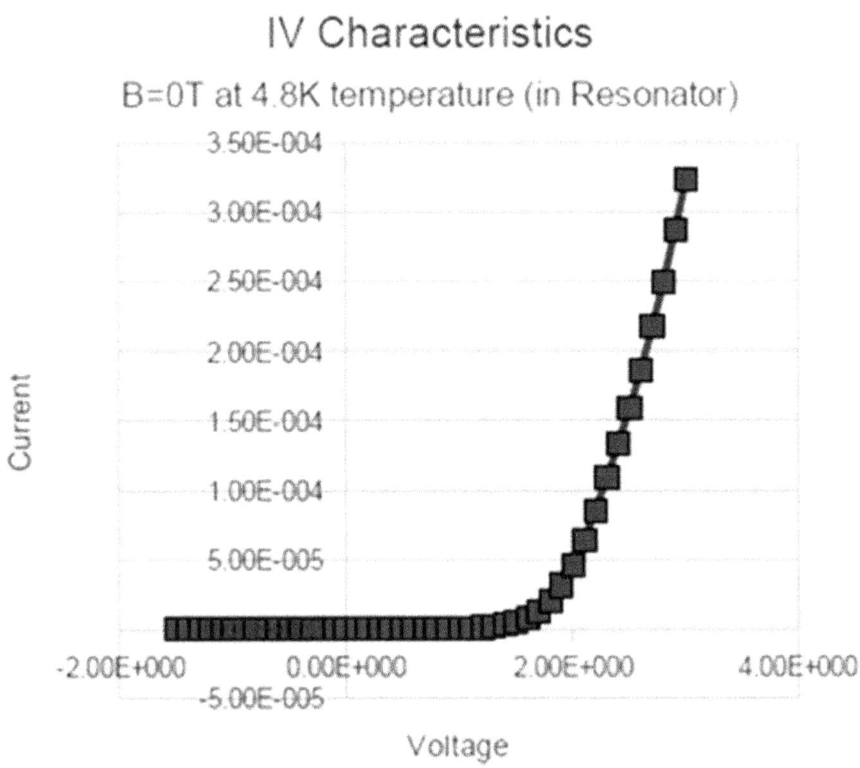

Optical Characterization and Measurements

To inject electrons from the spin aligner into the quantum dots, about 2.7V has been applied which changes the structure to near flat band conditions. Figure 12 shows IV characteristics for the mesa with a voltage supplied between -1.5 to 3V. Figure 13 is an electro-luminescence spectra of the mesa spin-LED sample without applied magnetic field.

Figure 13. EL spectrum

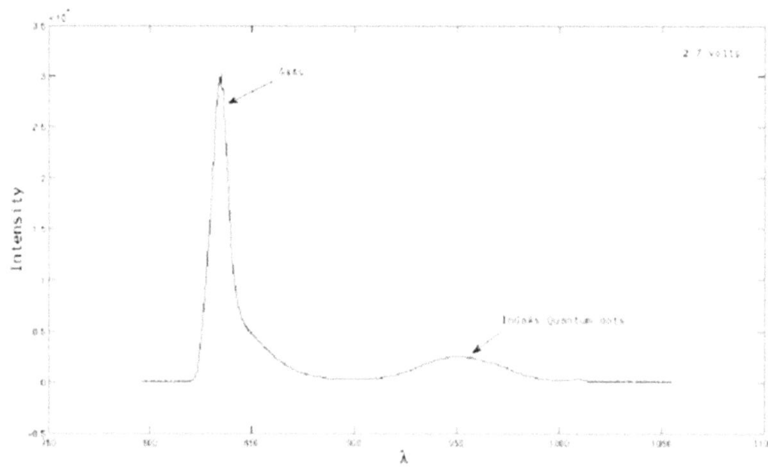

REFERENCES

Asshoff, P., Merz, A., Kalt, H., & Hetterich, M. (2011). A spintronic source of circularly polarized single photons. *Applied Physics Letters*, *98*(11), 112106. doi:10.1063/1.3564893

Datta, S., & Das, B. (1989). Electronic analog of the electrooptic modulator. *Applied Physics Letters*, *56*(7), 665–667. doi:10.1063/1.102730

Hettrich, M. (2010). *Electrical spin injection into single InGaAs quantum dots*. Institute of Applied Physics, KIT.

Ivo Timon, V. I. N. K. (2008). *Manipulation and Read-out of Spins in Quantum Dots*. Technische Universiteit Delft.

Loeffler, W. (2008). *Electrical preparation of spin-polarized electrons in semiconductor quantum dots*. Institute of Applied Physics, KIT.

Löffler, W., Höpcke, N., Kalt, H., Li, S. F., Grün, M., & Hetterich, M. (2010). Doping and optimal electron spin polarization in n-ZnMnSe for quantum-dot spin-injection light-emitting diodes. *Applied Physics Letters*, *96*(5), 052113. doi:10.1063/1.3308500

Merz, A. (2014). *Fast electron spin resonance controlled manipulation of spin injection into quantum dots*. KIT. doi:10.1063/1.4884016

Vitalii, Y. (2010). ODMR study of Zn1-xMnxSe/Zn1-yBeySe and (Cd1-x,Mn) Te/Cd1-yMgyTe Diluted Magnetic semiconductor quantum wells. *Applied Magnetic Resonance*, *39*(1-2), 31–47. doi:10.100700723-010-0133-0

Chapter 4
Electronic and Optical Properties of Quantum Nano-Structures:
Quantum Well Systems

Shivakumar Hunagund
https://orcid.org/0000-0001-5201-5827
Intel Corporation, USA

ABSTRACT

The aim of this chapter is to develop an understanding of the electronic and optical properties of quantum well systems. These structures, which can be found in semiconductors, confine particles in one dimension and exhibit discrete energy levels that can be calculated using fundamental quantum mechanics. Quantum wells are formed by sandwiching a material like GaAs between two layers of a material with a wider bandgap like AlAs, and can be grown using techniques such as MBE or CVD. The electronic and optical properties of quantum wells can be modified by altering parameters such as potential and well widths. In this chapter, the authors will use the Schrödinger equation to solve for the energy levels of quantum wells and provide a quantum mechanical description of the properties of electrons in these systems. They will also use computer programs to investigate the effects of changing parameters such as potential and well widths on the properties of quantum wells.

DOI: 10.4018/978-1-6684-7535-5.ch004

Copyright © 2023, IGI Global. Copying or distributing in print or electronic forms without written permission of IGI Global is prohibited.

INTRODUCTION

Quantum mechanics is the basis for our present understanding of physical phenomena on an atomic scale and has various applications in the field of engineering, lasers, semi-conductors and quantum optics. It gives the necessary tools with which to study and control objects on an atomic scale. Advancing technology of electronic devices necessitate shrinking the dimensions of size where single electron and quantum effects dominate the device performance and to determine their behavior, electronic and optical properties that are governed by the laws of quantum mechanics has to be investigated.

Rapid growth in planar growth techniques like molecular beam epitaxy (MBE) have made possible to grow high-quality, layered, semi-conductor hetero structures that gives exquisite control to achieve high precision monolayer growth within few nanometers and to make abrupt interfaces between semi-conductors with different bandgaps. Quantum wells, superlattices and heterojunctions were formed by growing arbitrary sequences of these layers. Electrons are confined at a heterojunctions to move in a thin 2D depletion layer just inside the low bandgap region. The laser that reads information from the discs is a quantum well laser that confines electrons to a thin, low bandgap 2D layer sandwiched between high bandgap layers material together forming the quantum well. The electronic properties of these structures can be controlled by the well size that could be used for the novel electronic and optical devices. Electron energies and thresholds for optical excitation can be customized, via

Figure 1. A 3D bulk structure, a heterojunction between high and low bandgap materials formed by layer by layer growth, a 2D QW formed by sandwiching a low bandgap layer between high bandgap regions and a QD formed by lateral confinement of the well in both dimensions

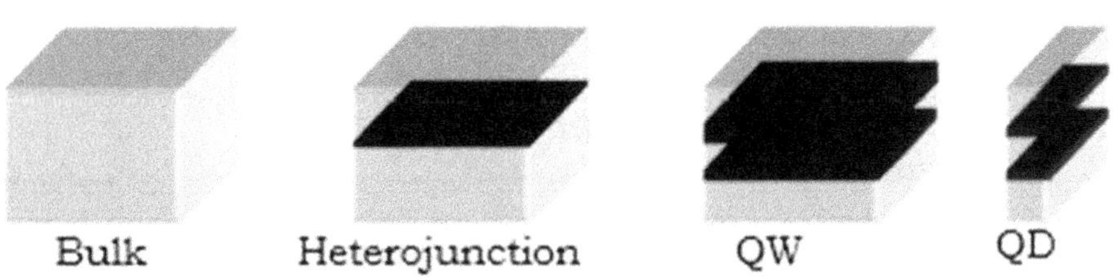

electron confinement, by changing the well layer thickness. Quantum well lasers, modulation field effect transistors etc., were some of the devices made by exploiting the quasi-2D electronic world.

A quantum well is a structure where the height is approximately the Bohr exciton radius while the length and breadth can be large, where as in quantum dot structure all dimensions are near the Bohr exciton radius. Investigating these structures with increasing quantum confinement (1D confinement for QW and 3D confinement for QD) achieved/influenced by different parameters of these structures is certainly intriguing and the optical devices made from the QDs are predicted to show improved performance over quantum wells. QDs find their applications in quantum information processing which requires structures that can be operated quantum mechanically for single quanta, such as, photons to transfer the information between individual quantum mechanical objects called qubits that spin electron confined in the dot or an optical excitation (exciton) localized to the dot.

For semi-conductor QWs and QDs, confinement leads to energy shifts and changes in the spatial overlap of conduction and valence electrons, which determine the strengths of optical transitions. Their electronic states and energy levels are determined by quantum mechanics. The structural size and shape can be used to quantum engineer their optical response by tailoring the desired transition energies, strengths and polarization dependence.

In this chapter we explore the electronic and optical properties of quantum well. Starting with the quantum well structures with confinement defined by the well then proceed to QDs of quantum dimensions in the proceeding chapter. Figures are created with Matlab programs to investigate the effects of changing parameters such as potential, well widths etc. for different types of wells, which may be used as a starting point in the development of designs for quantum mechanical devices.

SOLUTIONS TO SCHRÖDINGER'S EQUATION

Introduction to Schrödinger Equation

Electrons and holes in semiconductors are known to have both wave and particle like properties called wave – particle duality. The analysis treating free carriers as waves is referred to as the quantum approach and the other type of analysis as particles is called classical approach.

According to quantum mechanics, the behavior of a particle is described by the wave function Ψ (r,t), which is a solution of the Schrödinger wave equation. The wave function Ψ (r_1, r_2,....t) evolves in time according to the equation:

$$i\hbar\ \partial\Psi/\partial t = H\Psi \tag{1}$$

This is the Schrödinger equation that describes the space- and time-dependence of quantum mechanical systems. The operator H in the equation is the Hamiltonian operator for the system, the operator corresponding to the total energy.

$$H = T + U = -\hbar^2 / 2m * (d^2 / dx^2) + U(x) \tag{2}$$

where T and U are kinetic energy operator and potential energy operator for a particle of mass *m* moving in 1D. The Schrödinger equation (1D) can be separated into equations for the time and space variation of the wave function, when the potential energy is independent of the time.

$$\frac{-\Gamma^2}{2m}\frac{\partial^2\psi(x,t)}{\partial x^2} + U(x)\psi(x,t) = ih\frac{\partial\psi(x,t)}{\partial t} \tag{3}$$

$$-\hbar^2 / 2m\ (d^2\psi / dx^2) + U(x)\ \psi = E\ \psi \tag{4}$$

The *time-dependent Schrödinger equation* [equation 3] is a partial differential equation that determines the wave function at any other time from the known particle's wave function at t = 0. The state of motion of a particle is specified by giving the wave function. States in which the system has a definite total energy, the wave function is a standing wave and when the time-dependent Schrödinger equation is applied to these standing waves, it reduces to a simpler equation called the *time-independent Schrödinger equation* [equation 4] which allows us to find the wave functions of the standing waves and the corresponding allowed energy levels of quantum mechanical systems which are the eigen values of the wave equation. The associated wave function gives the probability of finding the particle at a certain position (refer appendix A).

The TISE, $\mathbf{H\psi = E\psi}$ (5) can be given in a matrix interpretation as

$$\sum_n H_{mn} c_n = E c_m \tag{6}$$

If m = n for a set of states then $H_{mn} = 1$ else $H_{mn} = 0$ and the hamiltonian has a diagonal matrix of the form $\mathbf{H}_{mm}\, \mathbf{c}_m = \mathbf{E}\, \mathbf{c}_m$ and the energy E is seen to be the diagonal element of the hamiltonian which implies solving the Schrödinger equation is equivalent to diagonalizing the hamiltonian matrix.

The Effective Mass Approximation

Majority of the semi-conductors are FCC Bravais lattice where the potential of the crystal consists of a 3D lattice of spherically symmetric ionic core potentials projected by the inner shell electrons those surrounded by the covalent bond charge distributions that stick everything together. In order to simplify this complex crystal potential, principle of simplicity is used where it is approximated as a constant such that Schrödinger equation can be applied. An empirical fitting parameter m^* (effective mass) is introduced since the crystal is not a vacuum. TISE then becomes:

$$-\hbar^2 / 2m^* (d^2 \psi / dx^2) = E \psi \tag{7}$$

$$\text{where } E = \hbar^2 k^2 / 2m^* \tag{8}$$

This effective mass approximation is widely used parameterization in semiconductor physics for e.g. GaAs has effective mass around 0.067 m_o.

Band Theory

Semiconductors consist of two distinct energy bands called valence band and conduction band. The lower band which originates from the valence electron states is responsible for the covalent bond formation and is almost full of electrons which can conduct by the movement of the empty states (holes). Electrons occupy energy levels from the lowest energies upwards. However, some energy levels are forbidden because of the wave like properties of atoms in the material. The allowed energy levels tend to form bands. The highest filled level at T=0 *K* is known

Figure 2. Energy (E) vs. wave vector (k) curves for an electron in the conduction band and a hole in the valence band of GaAs

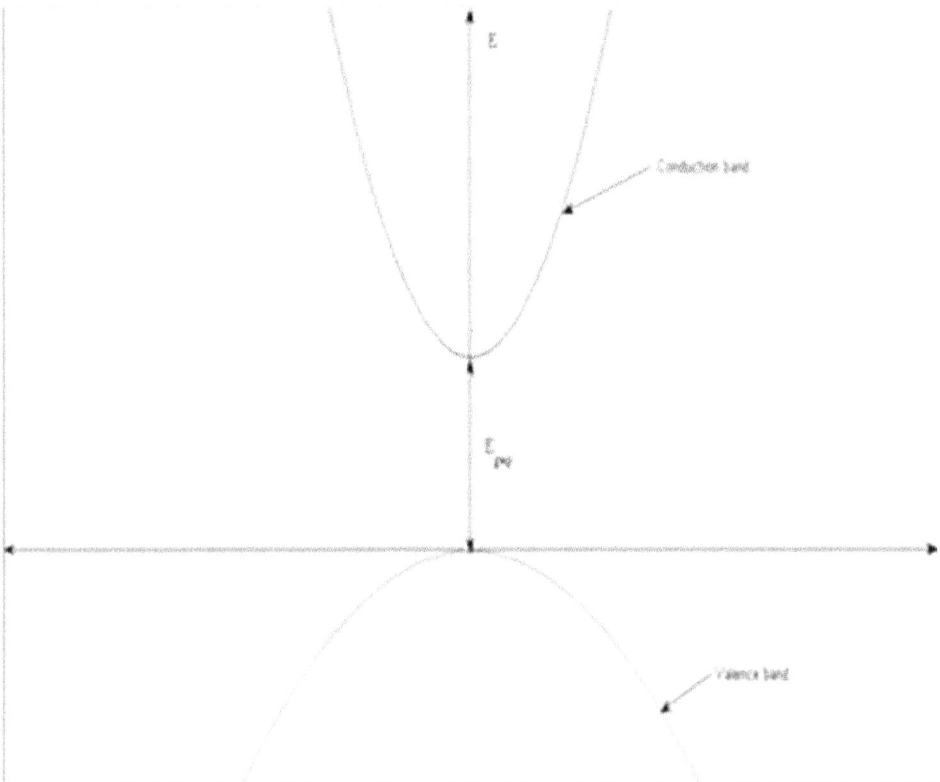

as the valence band. Electrons in the valence band do not participate in the conduction process. The first unfilled level above the valence band is known as the conduction band (represents excited electron states), which can be filled/accelerated by an applied electric field and therefore contribute to the current flow. The energy difference between the two bands is known as the bandgap E_g.

In compound semiconductors (GaAs, InP etc) the difference in electronegativity leads to a combination of covalent and ionic bonding. Ternary semiconductors are formed by the addition of a small quantity of a third element to the mixture, for example $Al_x Ga_{1-x} As$. The subscript x denotes the proportion elements that constitute the mixture of elements. Alloying semiconductors in this way allows the energy gap and lattice spacing of the crystal to be chosen to suit the application. This act of mixing different semiconductors with favorable properties into an alloy is usually referred to as bandgap engineering.

A large number of semiconductor materials can be used to manufacture quantum structures which unlocks fascinating possibilities for fabrication for e.g. laser structures: One can select semiconductor materials in such a way that the laser emits at an almost arbitrary wavelength. The best controlled III-V quantum system is the GaAs/AlGaAs structure with emission in the red range.

Heterojunctions and Heterostructures

When two semiconductors with different energy gaps are combined, a heterojunction is formed. The different combination level of doping and semiconductor material leads to different semiconductor junctions formation. Heterostructures are formed from multiple heterojunctions and are essential elements of the highest-performance optical sources and detectors. The usefulness of heterostructures is that they offer precise

Figure 3. A heterojunction between two different materials, shown here is the 1D potentials V(z) in the conduction and valence band

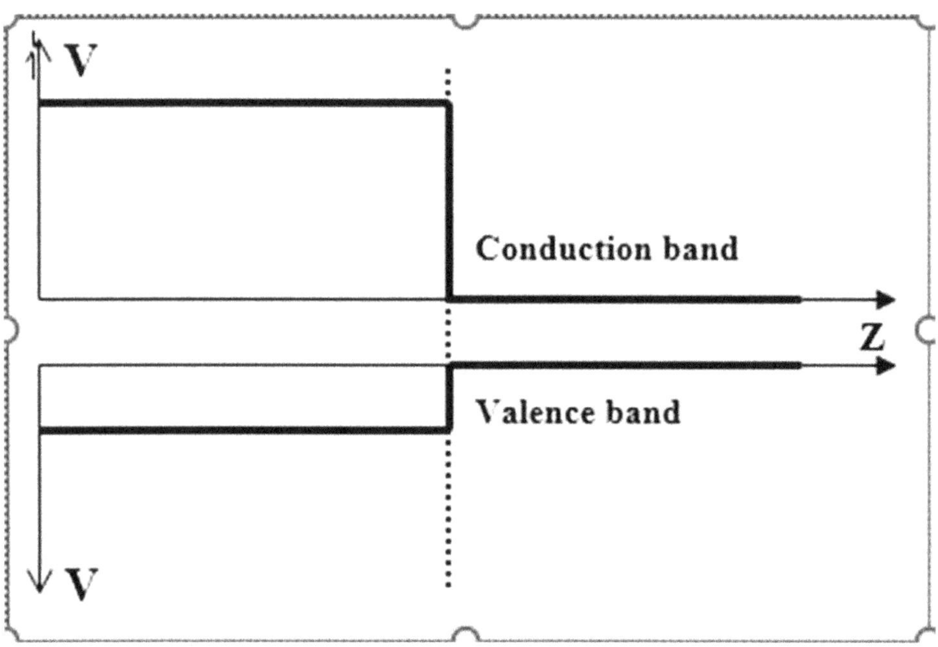

control over the states and motions of charge carriers in semiconductors. A heterostructure is defined as a semiconductor structure in which the chemical composition changes with position. The simplest heterostructure consists of a single heterojunction, which is an interface within a semiconductor crystal across which the chemical composition changes. Examples include junctions between GaSb and InAs semiconductors, junctions between GaAs and AlGaAs solid solutions, and junctions between Si and GeSi alloys. Most devices and experimental samples contain more than one heterojunction, and are thus more properly described by the more general term heterostructure.

The most important difference between these semiconductors in heterostructure devices is the energy gap and refractive index. Also, the threshold current density was improved at room temperature and it was achieved by cladding the active layer material, such as GaAs, with wider energy gap material, for instance $Al_x Ga_{1-x} As$ has very good lattice constant matching with GaAs.

A quantum well is formed between the barriers when a semiconductor material with a small energy gap is sandwiched between energy barriers from a semiconductor material with a larger energy gap. A typical layer thickness is in the range of 10-100 Ångström. When an electron is captured into this well, the probability to escape from the well is limited and due to the restriction on the movement of the electron into this plane (2D), affects the energy of the electron as compared to a free electron in the 3D case. Quantization effects will result in allowed energy bands, whose energy positions are dependent on the height and width of the barrier. The generalized Schrödinger equation for any one of the bands is given as:

$$-\hbar^2 / 2m^* (\partial^2 \psi(z)/ \partial z^2) + V(z)\, \psi(z) = E\, \psi(z) \quad (8)$$

(~the effective mass is taken to be same in each material)

Adding more semiconductor layers into the heterostructure results in the formation of stepped or asymmetric quantum well, as well as double or multiple quantum wells. Except in super lattice, multiple quantum wells exhibits the properties of a collection of isolated single quantum wells where they do not interact with each other. Structures having bandgap of one material lied within that of the wider bandgap material where charge carriers fall into quantum wells which are within the same layer of the material are called Type – I systems. In Type – II systems, the bandgap of the materials are aligned such that the quantum wells formed in the conduction and valence bands are in different materials where

Figure 4. A heterostructure shown with the 1D potentials V(z) in the conduction and valence band for a typical SQW and stepped QW

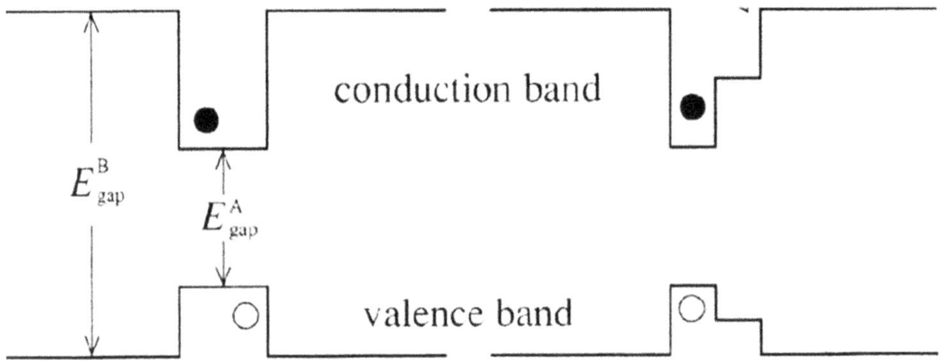

charge carriers are confined in different layers of the semiconductor. The recombination time of charge carriers is faster in Type – I systems.

The Envelope Function Approximation

Envelope function approximation is an approximation on the material, where it is derived from the deduction that physical properties can be derived from the slowly varying envelope function. In (equation 8), ψ (z) is identified as such rather than total wave function where it varies rapidly on the crystal lattice scale, (equation 8) for simple 1D Schrödinger equation represents the electronic structure. It has certain limitations that occur for very thin layers of material. The individual potentials in these thin layers become significant and the representation of crystal potential (bulk) as constant [*ref. section 2.2 The effective mass approximation*] breaks down. An alternative approach is made wherever such limitations occur.

The Infinite Well

For electrons bound between two infinitely high potential barriers (but free to move inside the well), the potential energy inside the well is U = 0 and rises abruptly to infinity at the edges. This type of system is called 1D square well or a particle in a box. The confinement of the particle causes energy to be quantized and the boundary conditions determine the allowed energies. The Schrödinger equation for this 1D box can be written as:

$$(d^2 \psi / dx^2) + (2m / \hbar^2) E \psi = 0 \quad (9)$$

This differential equation has general solution expressed as:

$$\psi (x) = A \sin kx + B \cos kx \quad (10)$$

Where $k = (2m_e E / \hbar^2)^{1/2}$ (11)

A and B are constants which can be determined by considering the boundary conditions:

$\psi (0) = 0$ and $\psi (L) = 0$

For $x = 0$, $\psi (0) = B = 0$, and $\psi (x = L)$:

$$\psi (L) = A \sin kL = 0 \quad (12)$$

which is satisfied only if kL is an integral multiple of π,i.e. $kL = n\pi$, where quantum number, n = 0, 1, 2 ....

From equation 8:

$$E_n = \hbar^2 \pi^2 n^2 / 2m_e L^2 \quad (13)$$

Therefore, the boundary condition quantizes the energy of the particle that is confined to a series of discrete values.

The only remaining unknown constant factor is A which is normalization constant. The probability of finding the particle somewhere within the box must be 1, so the integral of $\psi 2$ over the region between x = 0 and x = L must be equal to 1:

$$\int_0^L \psi * \psi \, dx = 1 \quad (14)$$

which gives $A = (2 / L)^{1/2}$

The complete solution is:

Figure 5. Schematic diagram of a ground state and two excited state energy levels and associated wave functions for an infinite deep square potential

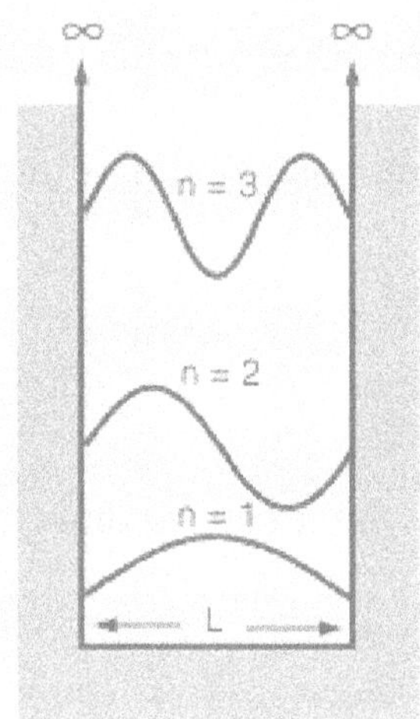

$$\psi_n = (2 / L)^{1/2} \sin (\pi n x / L) \tag{15}$$

$$E_n = h^2 n^2 / 8m_e L^2 \tag{16}$$

The squares of the wave function represent the probability densities for finding the particle in each state. As the value of n increases i.e. at higher state the particle is uniformly distributed. Another feature is that increasing number of nodes (point where ψ passes through zero) on rising n signify the higher energy state. The lowest energy that the particle can have i.e. at n = 1 is called the zero-point energy. The energy separation of neighboring states decreases as the walls move apart and give particle more freedom. As the length of the box approaches infinity, the separation of neighboring levels approaches zero, and the effect of quantization become completely negligible and the particle becomes unbounded and free. In other case smaller the confinement, the larger the energy.

If a particle is confined into a rectangular volume, the same kind of process can be applied to a 3D particle in a box, and the same kind of energy contribution is made from each dimension. Just as in 1D, where the wave functions look like those of a vibrating string with clamped

Figure 6. Energy levels of an infinite well

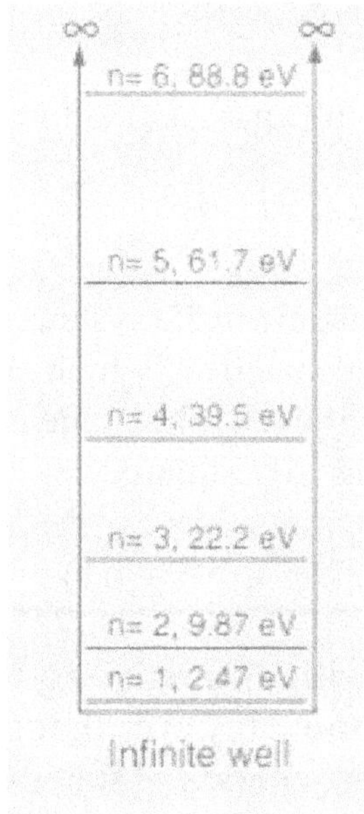

ends, so in 2D they can be expected to match to the vibrations of a plate with the edges rigidly fixed.

The hamiltonian for the 2D, infinitely deep square well is,

$$H = -\hbar^2 / 2m \, [(\partial^2 / \partial x^2) + (\partial^2 / \partial y^2)] + U(x,y) \tag{17}$$

$$U(x,y) = \begin{cases} 0 & \text{for } 0 \le x \le L_1 \text{ and } 0 \le y \le L_2 \\ \infty & \text{otherwise} \end{cases}$$

L_1 and L_2 are the linear dimensions along x and y direction. The Schrödinger equation for the particle inside the walls where the ψ is non-zero can be written as:

$$(\partial^2 \psi / \partial x^2) + (\partial^2 \psi / \partial x^2) = (2mE / \hbar^2)\, \psi \tag{18}$$

The boundary conditions are that the ψ must vanish at all the four walls. And the solution can be written as:

$$\psi_{n1,n2}(x,y) = 2 / (L_1L_2)^{1/2} \sin(\pi n_1 x / L_1) \sin(\pi n_2 y / L_2) \tag{19}$$

$$E_{n1,n2} = h^2 / 8m_e (n_1^2 / L_1^2 + n_2^2 / L_2^2) \tag{20}$$

To define the state of a particle in 2D system, the values of two quantum numbers (n_1 and n_2) must be specified which can take any integer values in their range independent of each other. One unique feature found in 2D is when the box is geometrically square, i.e.

$$L = L_1 = L_{2,} \rightarrow E_{n1,n2} = h^2 / 8m_e L(n_1^2 + n_2^2) \tag{21}$$

which implies that states with different quantum numbers for different systems has exactly the same energy even though their wave functions are different. This is known as degeneracy of states which is present in systems with high degree of symmetry. Degenerate eigen states correspond to identical eigen values of the Hamiltonian. Since eigen values correspond to roots of the characteristic equation. If the symmetry is broken by a perturbation, such as applying an external electric field, this can change the energies of the states, causing energy level splitting.

The Finite Well

The finite potential well (also known as the finite square well) is an extension of the infinite potential well, in which a particle is confined to a box. There is a probability associated with the particle being found outside the box which is unlike the infinite potential well. This contradicts the classical interpretation: if the total energy of the particle is less than potential energy barrier of the walls it cannot be found outside the box. In the quantum mechanical interpretation, there is a non-zero probability of the particle being found outside the box even when the energy of the particle is less than the potential energy barrier of the walls and this phenomena is known as quantum tunneling.

For electron bound in a potential well of finite depth U and width L, the potential energy inside the well (region II) is U = 0, and the Schrödinger equation for this region (same as equation 10) can be written as:

Figure 7. Schematic diagram of a ground state and two excited state energy levels and associated wave functions for a finite square potential well

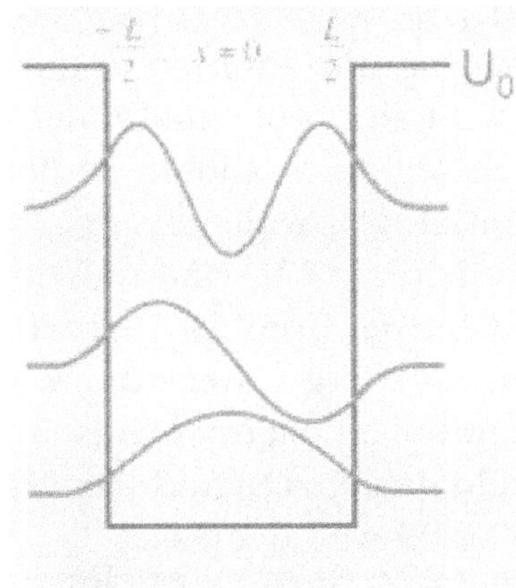

$$(d^2 \psi / dx^2) + (2m / \hbar^2) E \psi = 0$$

with the allowed wave functions $\psi(x) = A \sin kx + B \cos kx$ (same as [eqn. 2.0]). However the boundary conditions do not require for $\psi = 0$ at the walls, since there is a finite probability that electron may be found outside the wall (outside region II), and thus the ψ is generally non-zero in regions I and III. The Schrödinger equation for these regions can be written as:

$$(d^2 \psi / dx^2) - (2m (U - E) / \hbar^2) \psi = 0 \qquad (22)$$

where U > E. The general solution to the equation is,

$$\psi(x) = C\exp\{[2m (U - E)/ \hbar^2]^{1/2}x\} + D\exp\{-[2m (U - E)/ \hbar^2]^{1/2}x\} \qquad (23)$$

where C and D are constants that can be determined by applying boundary conditions. In region I ($x < 0$), the second term must be ruled out, (then $D = 0$) in order to avoid an infinite value for ψ for large negative values for x. In region III ($x > L$), the first term must be ruled out (then $C = 0$) in order to avoid an infinite value for ψ for large positive values of x. Therefore, for region I ($x < 0$), the solution is given as:

$$\psi_I(x) = C\exp\{[2m(U - E)/\hbar^2]^{1/2}x\} \quad (24)$$

and for region III (x > L), $\psi_{III}(x) = D\exp\{[2m(U - E)/\hbar^2]^{1/2}x\}$ (25)

These results indicate that the solution to the Schrodinger equation gives a wave function with an exponentially decaying penetration into the classically forbidden region. It should be noted that ψ is nonzero at the walls of the potential well which results in an increase of the de Broglie λ inside the well (region II) and thus lowering of the energy and momentum of the electron. Confining a particle to a smaller space requires larger confinement energy. Since the ψ penetration effectively enlarges the box, the finite well energy levels are lower than those for the infinite well. It should be noted that as the well width increases, the energy levels in the potential well decreases.

Also in finite potential well, potential is symmetric, the eigen states will have a definite symmetry: for symmetric (even parity), eigen states will be in cosine terms and the antisymmetric (odd parity) states as sine waves with the origin placed at the centre of the well.

Figure 8. Energy levels for an electron in a finite potential well of depth 64 eV and width 0.39 nm

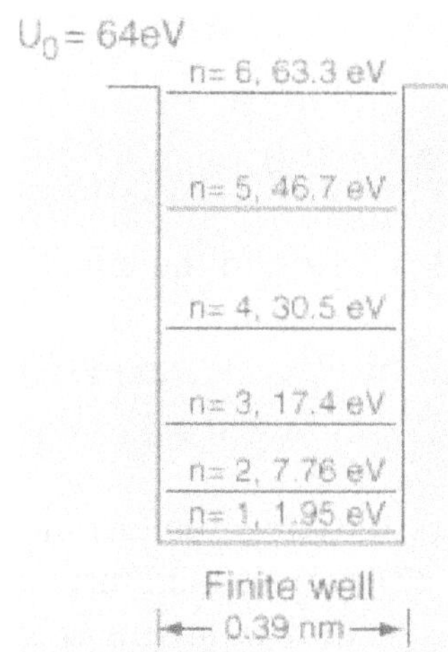

The Parabolic Potential Well

Harmonic oscillator is another example of a one-dimensional bound particle. The name simple harmonic oscillator is used in classical and quantum mechanics for a system that oscillates about a stable equilibrium point to which it is bound by a force obeying Hooke's law, $F = -kx$ where k is a force constant. The potential of the oscillator varies with displacement as:

$$U(x) = kx^2/2 \tag{26}$$

This parabolic potential is illustrated in fig. 2.7 where it is zero at $x = 0$. The difference between the parabolic potential and square well potential is the rapidity with which it rises to infinity. The walls of the oscillator are easily penetrable by the ψ to some extent. In harmonic oscillator, there is symmetrical occurrence of momentum and displacement in the expression for its total energy.

The hamiltonian operator for the harmonic oscillator is,

$$H = -\hbar^2 / 2m\ [(d^2 / dx^2) + kx^2/2] \tag{27}$$

The Schrödinger equation is therefore given as:

$$-\hbar^2 / 2m\ (d^2 \psi / dx^2) + kx^2/2 = E\ \psi \tag{28}$$

The energy of the harmonic oscillator is quantized and limited to the values:

Figure 9. Energy levels of the quantum harmonic oscillator

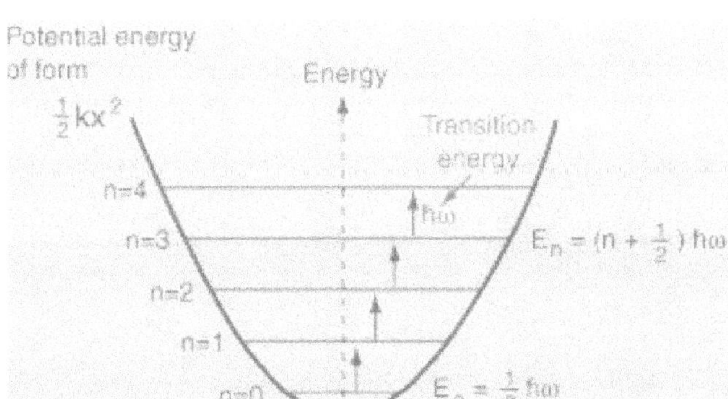

$$E_{\upsilon} = (\upsilon + ½)\, \hbar\omega \tag{29}$$

where $\omega=(k/m)^{1/2}$ $\upsilon = 0, 1,2\ldots$

The ψ for the oscillator has a form that resembles Gaussian function, a function of the form e^{-x2} multiplied by the polynomial in the displacement:

$$\Psi_{\upsilon} = N_{\upsilon}\, H_{\upsilon}\, (y)\, e^{-1/2y2}$$

$$y = x \,/\, \alpha \quad \alpha = (\hbar^2 \,/\, mk)^{1/4} \tag{30}$$

Where N_{υ} is a normalization constant: $N_{\upsilon} = (1/\, 2^{\,\upsilon}\, \upsilon!\pi^{1/2}\, \alpha)^{1/2}$

The significant point about the energy levels is that they form a ladder with equal spacing. The energy separation between neighbors is $E_{\upsilon+1} - E_{\upsilon} = \hbar\omega$ regardless of the value of υ. As the force constant k increases so the separation between the neighboring levels also increases (since $\omega \propto k^{1/2}$). As k decreases or mass increases, so ω decreases, and the separation between neighboring levels decreases. From equation 28, the greater the force constant, the more confining the potential and therefore the more sharply curved the wave functions.

Figure 10. The first three energy levels and wave functions of the simple harmonic oscillator.

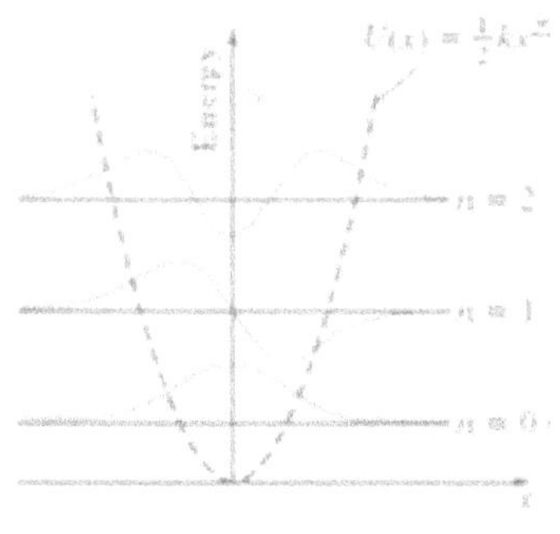

Addition of Electric Field

A linear potential is added for an electron of charge -e in the conduction band (-eFz) when an electric field is applied along the growth (z-). The ground state energy level with the electric field along the growth axis as the perturbation is given as:

$$\Delta E^{(1)} = \int_{-\infty}^{+\infty} \psi_1 {}^*(z)(-eFz)\, \psi_1(z)dz \tag{31}$$

The Schrödinger equation with an applied electric field of strength is given as:

$$\partial^2 \psi / \partial z'^2 - \mathbf{z'}\, \psi = 0 \tag{32}$$

$$z' = (2m^*/\hbar^2)^{1/3} [\{(U(z) - E) / (eF)^{2/3}\} - (eF)^{1/3} z]$$

The standard solution which is a linear combination of Airy functions:

$$\psi(z') = A\, Ai(z') + B\, Bi(z') \tag{32}$$

Under the influence of large electric field, a charged particle prefers to move to areas of lower potential where the electron within the quantum well moves to the right hand side of the well lowering its total energy. This phenomenon is commonly observed in heterostructures and suppression of the confined energy level by an electric field is known as 'quantum stark effect'.

Applications

Quantum well heterostructures (QWHS) are a type of semiconductor device that consists of layers of different semiconductor materials stacked on top of each other. These layers, or wells, are typically only a few nanometers thick that are formed by creating a series of thin layers of different semiconductor materials and are separated by thin layers of barrier material. The unique properties of these heterostructures arise from the quantum confinement of electrons and holes in these structures which are confined in two dimensions, and their energy levels can be precisely controlled by changing the thickness and composition of the layers, which can lead to a range of interesting electronic, optical, and magnetic phenomena. QWHS are key components of many optoelectronic devices, because they can increase the strength of electro-optical interactions by confining the carriers to small regions and due to their quasi-2D nature, electrons in quantum wells have a sharper density of states than bulk materials.

As the active layer thickness in a double heterostructure becomes close to the de-Broglie wavelength quantum effects become apparent. Quantum wells are important in semiconductor lasers (e.g. active regions of laser diodes or surface-emitting) because they allow some degree of freedom in the design of the emitted wavelength through adjustment of the energy levels within the well by careful consideration of the well width and thereby allow reaching significantly lower pump thresholds than with thicker layers.

QWHS have been a subject of intense research in the field of condensed matter physics due to their unique electronic, magnetic and optical properties. This has led to a wide range of applications, including high-speed electronic devices, lasers, and photodetectors.

Quantum wells are also used as absorbers in semiconductor saturable absorber mirrors (SESAMs), and in electro-absorption modulators. If a large amount of optical gain or absorption is required, MQWs can be used, with a spacing typically chosen large enough to avoid overlap of the corresponding wave functions.

Ultra short laser pulses using passive mode-locking is being achieved through independent control in the gain media and cavity loss, where laser mode-locking is achieved with a MBE structure with multiple quantum wells (Quintero-Torres, 2005).

Some of the other examples of applications include: cascade infrared laser, multi-color and tunable infrared detectors, as well as the focus plane array based on the intersubband or interband transitions in usual multi quantum well or type-II superlattice structures. Quantum well infrared photodetectors are based on quantum wells, and are used for infrared imaging (Shen, 2000).

Researchers at US microchip company Intel and UK research firm Qinetiq developed a transistor that uses one-tenth of the energy of existing components, using a novel semiconducting material - indium antimonide.

Transistors act as switches or amplifiers in electronic circuits to process information. Indium antimonide allows electrons to speed through faster than conventional silicon-based transistors due to its highly active and greater number of charge carriers that help relay the electrons quickly. But these charge-carriers also make these transistors more difficult to control than silicon ones, except at extremely low temperatures - around 77 Kelvin (-196°C). To overcome this temperature limitation, the researchers sandwiched pure indium antimonide between layers of the same material mixed with aluminium. The isolated pure material acts as a quantum well, confining electrons which travel at high speed but which can also be controlled at very low voltage (Knight, 2005).

In recent years, research in quantum well heterostructures has been focused on developing new device architectures and materials combinations to enhance their performance and expand their range of applications. One area of particular interest is the development of novel optoelectronic devices, such as lasers and photodetectors, which can operate at higher frequencies and longer wavelengths compared to traditional semiconductor devices. Other research has focused on the use of quantum well heterostructures in energy harvesting and storage applications, such as photovoltaic cells and lithium-ion batteries. Additionally, there has been significant research in the use of quantum well heterostructures for spintronic applications, where the spin of the electrons is exploited for information processing and storage.

However, there are several challenges and limitations in this research. One of the main challenges is the fabrication of high-quality QWHS. The thin layers in these structures need to be precisely controlled in order to achieve the desired properties. This requires advanced fabrication techniques, such as molecular beam epitaxy and metal-organic chemical vapor deposition. These techniques are complex and time-consuming, and they can be prone to defects and impurities.

Another challenge is the lack of a comprehensive understanding of the electronic and optical properties of QWHS. These properties are determined by the quantum mechanical behavior of the electrons, which can be difficult to predict and measure. There is a need for more theoretical and experimental studies to fully understand the behavior of these systems.

In addition, there are limitations in the materials that can be used to create QWHS. Currently, most QWHS are made from traditional semiconductor materials, such as silicon and germanium. While these materials have good electronic and optical properties, they are not ideal for all applications. For example, silicon-based QWHS are not efficient for optical applications due to their high absorption coefficient. There is a need for the development of new materials and techniques that can overcome these limitations.

Overall, the current state of research in quantum well heterostructures is very active and there are many exciting developments on the horizon. These materials and devices hold great potential for a wide range of applications in fields such as telecommunications, renewable energy, and computing, and are likely to continue to be an important focus of research in the coming years.

REFERENCES

Atkins, P. W. & Friedman, R. S. (2010). *Molecular Quantum Mechanics.* Oxford university press, UK.

Brian L.F. (2004). *Matlab tutor CD.* John Wiley & Sons, (UK).

Bryant G.W. & Solomon G.S. (2004). *Optics of Quantum dots and wires.* Artech house, Inc. MA, US.

Van Loan, C. (1996). *Introduction to scientific computation – A Matrix vector approach using Matlab.* Prentice hall, NJ

Cleve B. Moler. (2004). *Numerical Computing with Matlab.* Siam Publishing, Philadelphia.

Hanselman, D & Littlefield, B. (2000). *Mastering Matlab 6 – A Comprehensive tutorial and reference.* Prentice hall, NJ.

Harrison, P. (1988). Quantum Wells, Wires and Dots: Theoretical and Computational Physics. John Wiley & Sons, (UK).

Harrison, P. (1997). *Computational methods in Physics, Chemistry & Biology: An Introduction.* John Wiley & Sons, (UK).

Jacak, L. & Wojs, H. (1998). *Quantum dots.* Springer, Berlin.

Thijssen, J. (2007). *Computational Physics (2nd edn.).* Cambridge university press, UK.

Knight, W. (2005, February). 'Quantum well' transistor promises lean computing. *New Scientist*, 10.

Levi, A. (2012). *Applied Quantum Mechanics (2nd edn.).* Cambridge university press, UK.

Michler, P. (2003). *Single quantum dots: Fundamentals, applications and new concepts.* Springer, Berlin.

Pratap, R. (2009). *Getting started with Matlab – A quick introduction for Scientists and Engineers.* Oxford university press, UK.

Quintero-Torres, R., Vázquez-Cerón, E., Rodríguez-Rodríguez, E., Stintz, A. and Diels, J.-C. (2005). Multiple quantum wells for passive ultra short laser pulse generation. phys. stat. sol. (c), 2: 3015-3018.

Liboff, R. (2008). *Introductory Quantum mechanics (4th edn.).* Pearson education Inc.

Vukmirovic, N., Gacevic, Z., Ikonic, Z., Injin, D., Harrison, P., & Milanovic, V. (2006, July). Intraband absorption in InAs/GaAs quantum dot infrared photodetectors –effective mass versus k x p modelling. *Semiconductor Science and Technology*, 3.

Weiss, P. (2006). Quantum-Dot Leap. *Science News Week, 169*(22), 344.

Xie, Z. G., Götzinger, S., Fang, W., Cao, H., & Solomon, G. S. (2007). Influence of a single quantum dot state on the characteristics of a microdisk laser. *Physical Review Letters*, *98*(11), 117401. doi:10.1103/PhysRevLett.98.117401 PMID:17501091

Chapter 5
Quantum Engineering:
Quantum Dots

Shivakumar Hunagund
https://orcid.org/0000-0001-5201-5827
Intel Corporation, USA

ABSTRACT

Quantum dots are semiconductor nanostructures that exhibit unique electronic and optical properties. These structures are formed through lithographic patterning of quantum wells or spontaneous growth techniques, and can be considered as 3D quantum wells with no degrees of freedom and quantized levels for all three directions of motion. Due to their small size, quantum dots have dimensions that are similar to the Bohr exciton radius, and they confine electrons in all three spatial directions to produce discrete energy levels that are similar to those found in atoms. One of the key features of quantum dots is that their electronic and optical properties can be tailored by altering parameters such as size and shape. In this chapter, the authors will use the Schrödinger equation to solve for the energy levels of different types of quantum dots, including infinite (1D and 2D) and finite potential wells. They will also use Matlab programs to solve for a realistic model of a quantum dot, investigating the effects of changing parameters such as potential and well widths.

DOI: 10.4018/978-1-6684-7535-5.ch005

Copyright © 2023, IGI Global. Copying or distributing in print or electronic forms without written permission of IGI Global is prohibited.

1. INTRODUCTION

Quantum dots are semiconductor nanostructures that exhibit unique electronic and optical properties. These structures are formed through lithographic patterning of quantum wells or spontaneous growth techniques, and can be considered as 3D quantum wells with no degrees of freedom and quantized levels for all three directions of motion. Due to their small size, quantum dots have dimensions that are similar to the Bohr exciton radius, and they confine electrons in all three spatial directions to produce discrete energy levels that are similar to those found in atoms.

One of the key features of quantum dots is that their electronic and optical properties can be tailored by altering parameters such as size and shape. In this chapter, we will use the Schrödinger equation to solve for the energy levels of different types of quantum dots, including infinite (1D and 2D) and finite potential wells. We will also use Matlab programs to solve for a realistic model of a quantum dot, investigating the effects of changing parameters such as potential and well widths.

The Schrödinger equation, which describes the wave-like nature of an electron and defines its motion, is a fundamental tool for understanding the behavior of quantum systems. By solving this equation for different types of quantum dots, we can gain a deeper understanding of the electronic and optical properties of these systems and how they can be controlled and manipulated. In this chapter, we will provide qualitative descriptions of each case and explore the ways in which the properties of quantum dots can be modified and tailored to meet specific needs. Overall, this chapter will provide a comprehensive introduction to the exciting field of quantum dots and the many ways in which these structures are being used in a variety of applications.

Quantum confinement is when electrons and holes in a semiconductor are restricted in one or more dimensions and occurs at dimensions of a nanocrystal that is comparable to the size of an exciton in bulk crystal, called the Bohr exciton radius. A quantum dot is a semiconductor nanostructure and confines the charge carriers in all three dimensions.

Its structure may be considered as a 3D quantum well without any degrees of freedom and therefore has 0D properties with quantized levels for all three directions of motion. For rectangular dot with dimensions (L_x,L_y,L_z), the energy levels of the system is given as:

$$E\,(n_x, n_y, n_z) = \hbar^2\pi^2/2m^*\,(n_x^{\,2}/L_x^{\,2} + n_y^{\,2}/L_y^{\,2} + n_z^{\,2}/L_z^{\,2}) \tag{1}$$

where the quantum numbers specify the quantized levels in each direction. It is assumed here that all three directions have infinite barriers. The energy spectrum of the structures is therefore discrete which is similar to the energy spectrum of the atoms. But the greatest advantage here lies in tuning the position of the energy levels by altering the size of the quantum dot. By confining the charge carriers in all 3D, increases the electron – hole overlap and thus increase the radiative quantum efficiency.

Figure 1. Fluorescence induced by exposure to UV light in vials containing various sized Cadmium selenide (CdSe) quantum dots

Since the electrical conductivity of semiconductors can be greatly altered via an external stimulus therefore they form critical parts of many different kinds of electrical circuits and optical applications, furthermore the Quantum dots are so small, ranging from 2-10 nanometers (10-50 atoms) in diameter, they behave differently making them suitable for applications in areas where it hasn't been possible before. The confinement of electrons in the quantum dots can be due to electrostatic potentials, presence of an interface between different semiconductor materials, semiconductor surface, or due to a combination of these. This localization of electrons produces discrete electron energy levels spectrum similar to atoms. The corresponding wave functions are spatially localized within the quantum dot, but extend over many periods of the crystal lattice. A quantum dot contains a small finite number (of the order of 1-100) of conduction band electrons, valence band holes, or excitons. The atomic kind of behavior of quantum dots thus gives an opportunity to engineer the electronic and optical components (energy spectrum) outside those naturally existing 92 elements by controlling the geometrical size, shape, and the strength of the confinement potential. The properties of electrons confined in these zero dimensional quantum dots can be described by quantum theory. The ability to tailor the quantum dots to specific optical and electronic properties, small number of electrons confinement, facilitates carrying out ab initio calculations and understanding of some of the effects like quantum hall effect which is characteristic to such electronic systems through numerically solving the Schrodinger equation.

2. FABRICATION

Self-Organization

Self-assembled quantum dots nucleate spontaneously under certain conditions (temperature, growth rate, mode of deposition etc.) during molecular beam epitaxy (MBE) and metallorganic vapor phase epitaxy (MOVPE), when a material is grown on a substrate to which it is not lattice matched. The resulting strain is exploited to obtain quantum dots. This growth mode is known as Stranski-Krastanov growth. The quantum dot island is then capped with a layer of the substrate material. This fabrication method has potential for applications in quantum cryptography (i.e. single photon sources) and quantum computation. Size and uniformity can be enhanced by forming multilayers. The main limitations of this method are the cost of fabrication and the lack of control over positioning of individual dots.

Lithographic Technique

There are many different lithographic techniques (X-ray lithography, electron and focused ion beam, scanning tunneling lithography, optical lithography and holography etc.) in practice to produce QDs. The first technique employed to achieve 3D confinement of the carriers in semiconductor heterostructures were lithographic patterning and etching of quantum well structures. Individual quantum dots can be created from two-dimensional electron or hole gases present in remotely doped quantum wells or semiconductor heterostructures. The sample surface is coated with a thin layer of resist. A lateral pattern is then defined in the resist by electron beam lithography. This pattern can then be transferred to the electron or hole gas by etching, or by depositing metal electrodes (lift-off process) that allow the application of external voltages between the electron gas and the electrodes. Such quantum dots are mainly of interest for experiments and applications involving electron or hole transport (electrical current). To achieve higher-resolution, SEM is employed that can draw fine scale of patterns.

3. ELECTRONIC AND OPTICAL PROPERTIES

The energy levels of quantum dots can be probed by optical spectroscopy techniques similar to atoms. The expected discrete electronic energy level shell structure was first observed using capacitance (C-V) and IR spectroscopy measurements on large ensemble of InAs quantum dots embedded into a GaAs/AlGaAs MIS device structure. The number of electrons which can be confined to the quantum dots can be controlled by their dimensions. The techniques of capacitance spectroscopy are based on the measurement of the electrical capacity of the system $C = dQ / dV$ (that is directly related to the density of states $dN / d\mu$) as a function of the gate voltage V applied perpendicular to the quantum well. The quantum well in which the quantum dots are created serves like storage of free carriers and is placed above the doped layer electrode. Altering the voltage leads to change in number of electrons transferred from the well to the dot. The addition of electron which is dependant on the chemical potential of the quantum dot occurs only at discrete values of V due to the discrete spectrum of energy levels in quantum dots.

Figure 4 shows C-V spectrum for a diode containing InAs quantum dots with two series of peaks associated respectively with the *s* and *p* shell loading when electrons are tunneling from the back n^+ gate as a positive voltage is applied to the front one. The width of the peaks is dependent upon the quantum dot size dispersion and their numbers in the measuring device. It can be clearly resolved from fig. that the range of voltage V where the dot is empty, a strong peak corresponds to the transfer of first electron to the dot and a series of weaker peaks correspond to capturing consecutive electrons in the dot. The total energy of the system increases with decreasing dot size and the distance between subsequent peaks. Under the influence of magnetic field results in rearrangement of energy levels in the dots, changes the intensity of peaks and energy shift is observed.

Figure 2. Capacitance spectra of QDs in different magnetic fields
Source: Jacak and Wojs (1998)

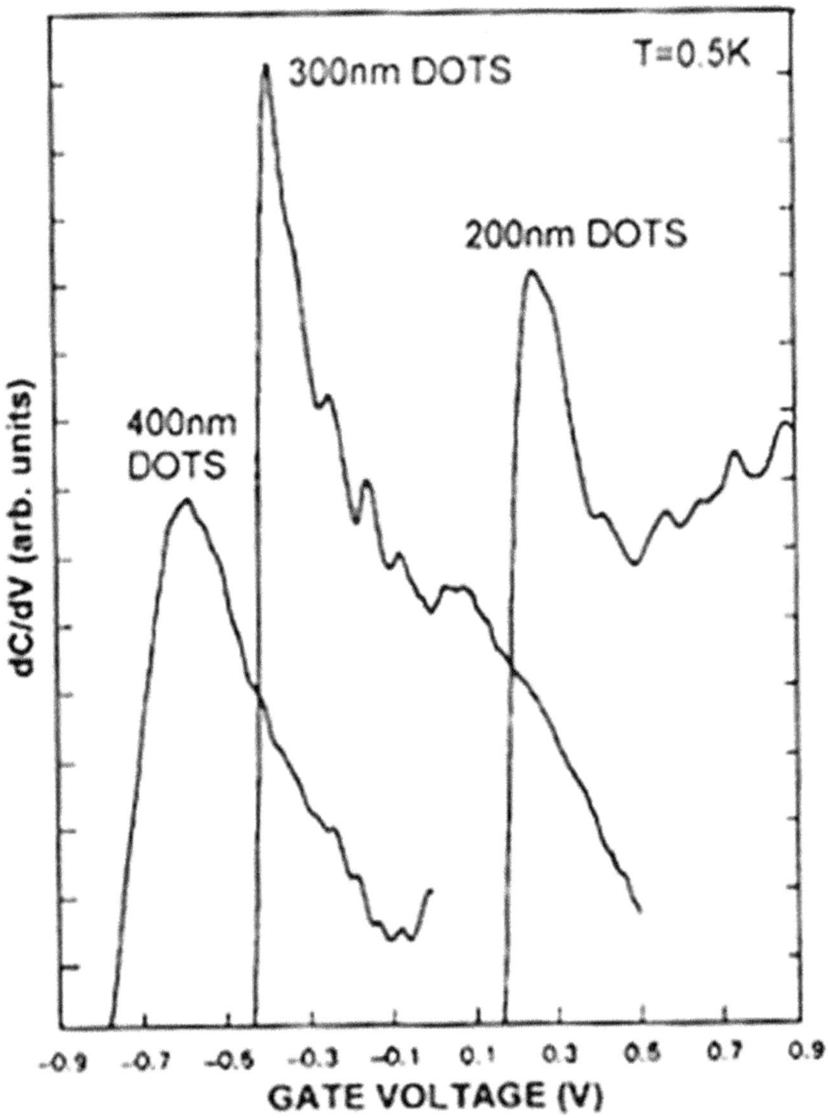

Figure 3. Capacitance spectra of QDs for different diameters
Source: Michler (2003)

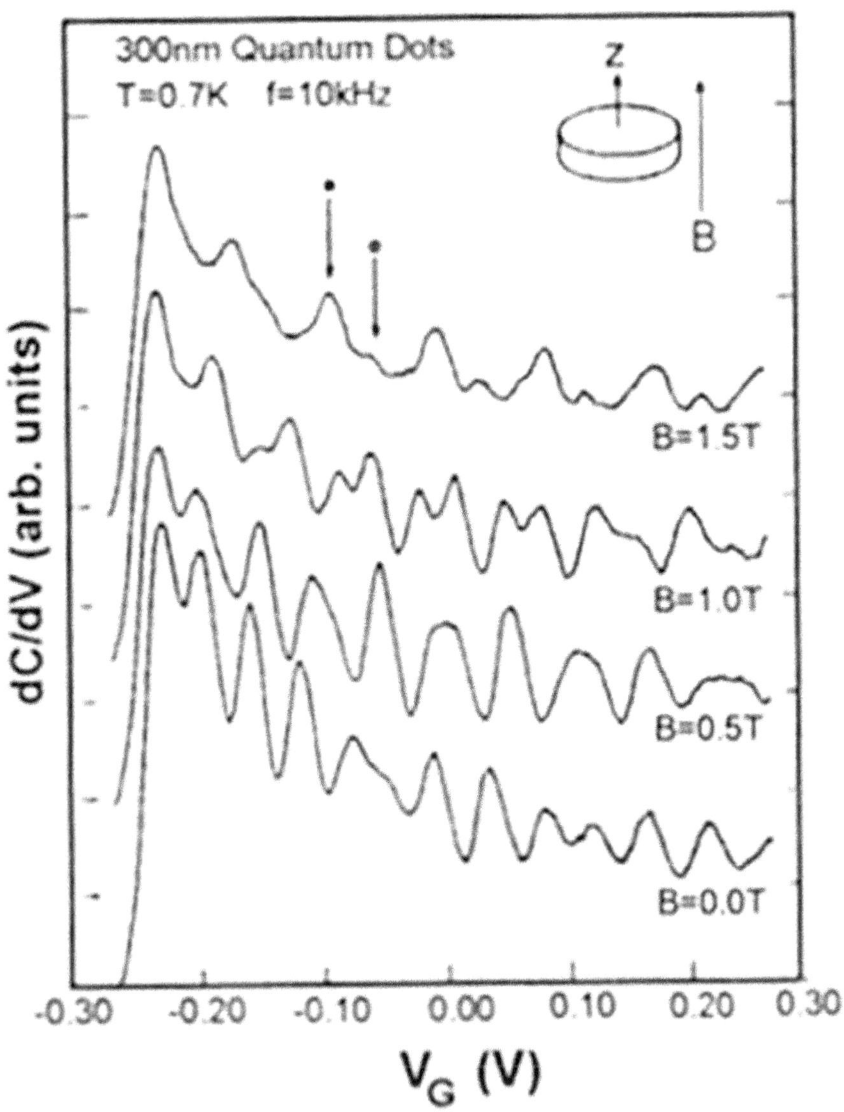

Figure 4. Capacitance spectra of QDs for CB diagram showing loading of s and p shell electrons
Source: Michler (2003)

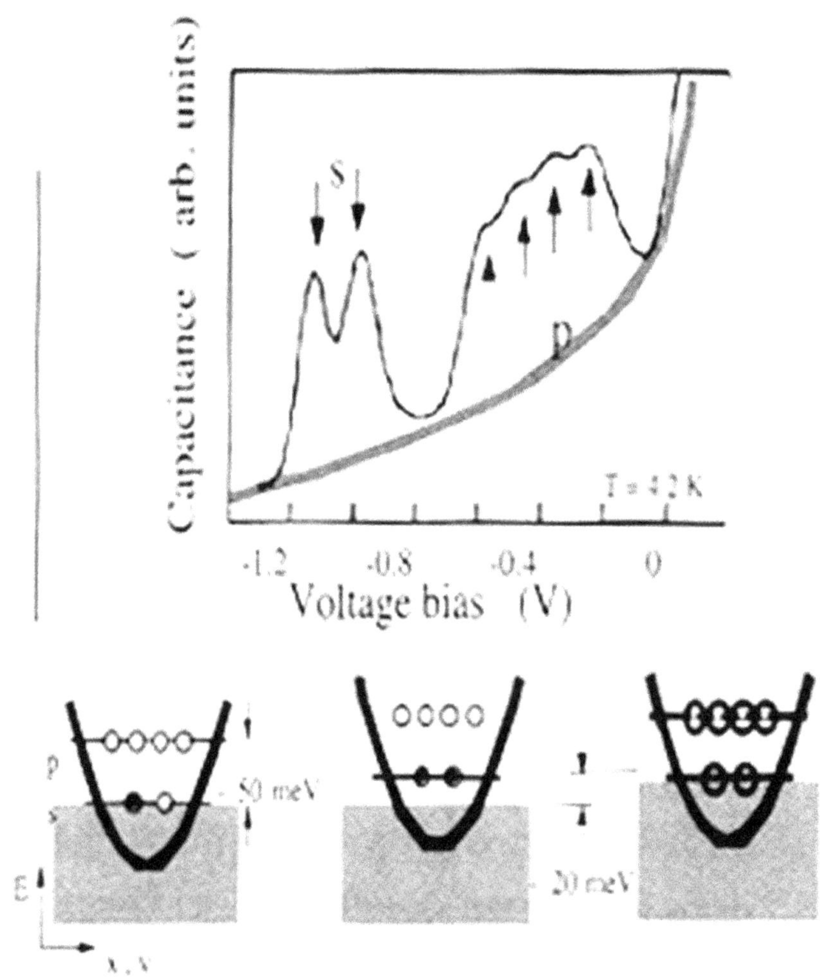

The electron energy levels in quantum dot are discrete, the addition or removal of atoms to the quantum dot has the effect of altering the boundaries of the bandgap. Altering the geometry of the surface of the quantum dot also changes the bandgap energy, owing again to the small size of the dot, and the effects of quantum confinement. The confinement of electrons and holes in the quantum dots is blue shifted at interband absorption edge since the bandgap in quantum dots are always energetically larger, as a result quantum dots of the same material, but with different sizes, can emit light of different colors.

Increasing the dot makes it appear redder i.e. the more towards the red end of the spectrum and hence the fluorescence and decreasing the size results in blue i.e. the more towards the blue end. The coloration is directly related to the energy levels of the quantum dot. The bandgap energy that determines the energy (and hence the color) of the fluoresced light is inversely proportional to the square of thc sizc of thc quantum dot.

Figure 5. Photoluminescence spectra of single quantum dots
Source: Jacak and Wojs (1998)

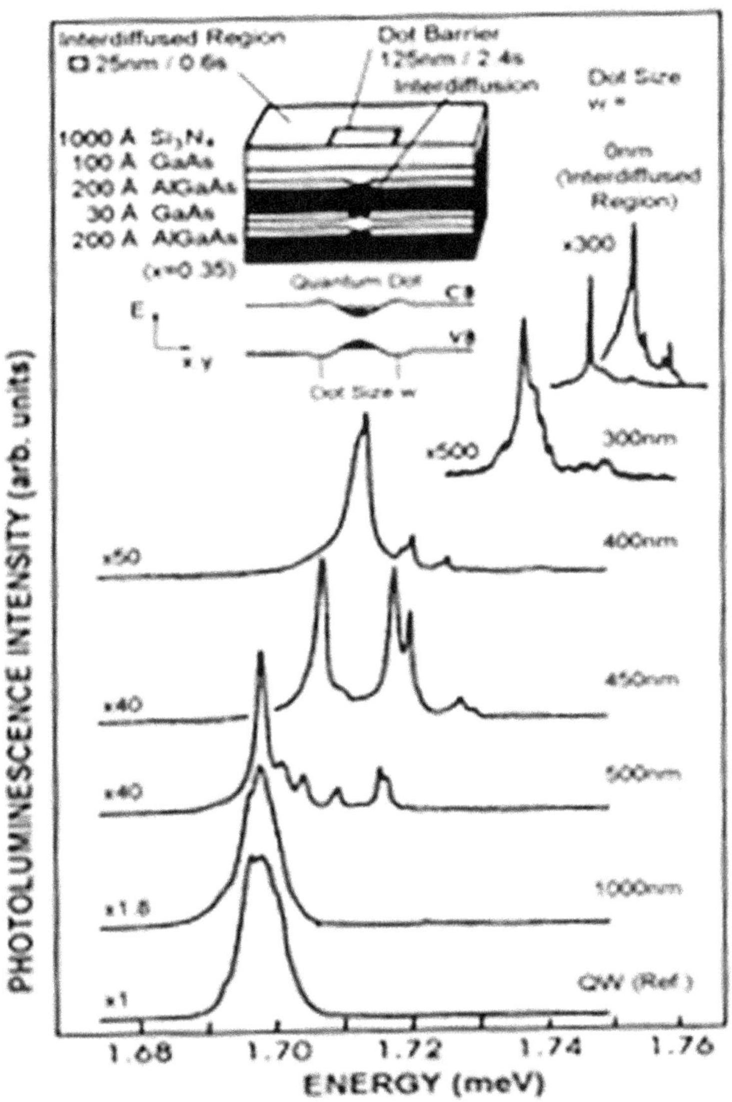

4. QUANTUM BOXES

Consider the case for the cuboid quantum dot with an infinite potential separating the inside of the box from the outside, the 3D Schrödinger equation within the box is given as:

Figure 6. Schematic illustration of a quantum box with side Lx, Ly and Lz at x,y and z axis

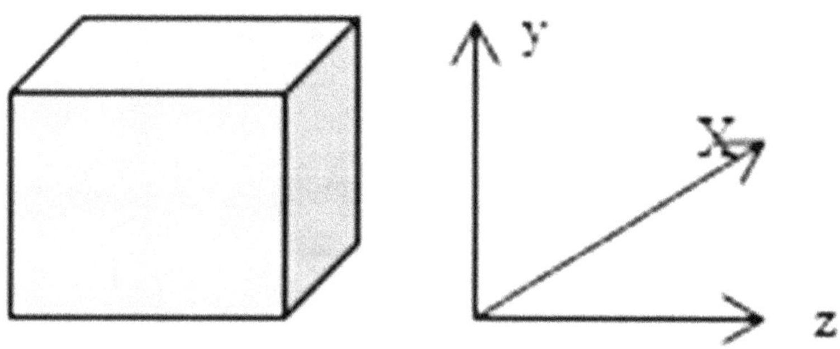

$$-\hbar^2 / 2m^* [(\P^2 / \P x^2) + (\P^2 / \P y^2) + (\P^2 / \P z^2)]\ \psi(x,y,z) = E_{x,y,z}\ \psi(x,y,z) \quad (2)$$

The confinement energy within this box is given by the eqn. 4.4. A 3D solution can be constructed by expanding the ψ as a linear combination of infinite well solutions (Harrison, 2016).

5. EFFECTIVE MASS MODEL

Quantum dot infrared photodetectors comprising III – As self-assembled QDs is used for detecting mid- and far- infrared EM radiation. Quantum dot – in – a – well photodetectors provide a way to tune the detection wavelength within a certain range by changing the well width and focal plane arrays based on QDIPs. The energy levels and the wave functions can be calculated for the Qdot using the effective mass model approach which is simpler and faster compared to *k* x *p* method (*k* x *p* is more realistic method) (Vukmirovic et al., 2006).

Consider a quantum dot which has a cylindrical symmetry shape. Most QDIPs have the shape of the lens, cone or truncated cone which all belong to the class of cylindrical symmetric dots. Within the framework of the effective mass method, the hamiltonian is given by,

$$\hat{I}I = -k(\hbar^{2} / 2m^{*}(r))k + U\,(r) + |e|\,Fz \qquad (3)$$

where k_i (iÎ{1,2,3}) is the differential operator k_i = -i¶ / ¶x_i, F = Fe_z is the electric field oriented along the z- direction, m*(r) is the position dependent effective mass and U(r) is the position dependent potential, both are assumed constant within the dot and the matrix. m* is taken as $0.04m_0$ and the conduction band offset U_o = 450meV are used to take the averaged effect of strain into account.

Figure 7. Schematic illustration of a QD in an embedding cylinder of radius R_t and height Ht. The wetting layer width is d_{wl}.

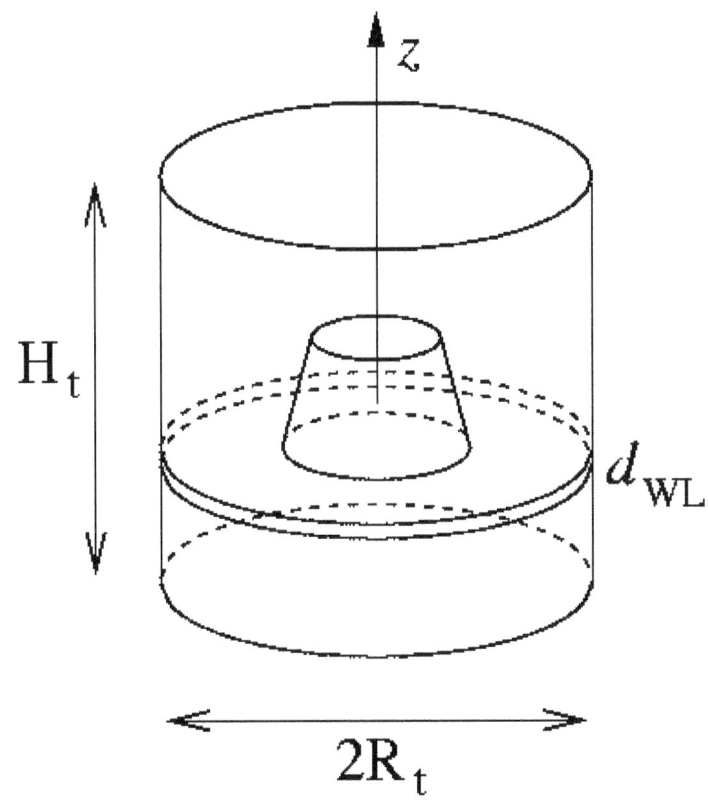

To find the eigen energies and the corresponding wave functions, orthonormal wave function expansion method was used, which is based on embedding the dot in a cylinder of radius R_t and height H_t. The wave function is assumed as a linear combination of the expansion basis functions,

$$\psi (r) = A_{nl} b_{nml} (r, z, \varphi) \quad (4)$$

with the co-efficients A_{nl} ($n \hat{I} \{1 \ldots n_{max}\}$) and $l \hat{I} \{-l_{max} \ldots .. l_{max}\}$ to be determined. Given l and m are integers and n is a positive integer (Vukmirovic et al., 2006) for basis functions equations. The Hamiltonian matrix for the eigen value problem is given as:

$$H_{nml,n'ml'} A_{n'l'} = E A_{nl'} \quad (5)$$

$$\hat{H}_{nml,n'm'l'} = b^*_{nml} \hat{H} b_{n'm'l'} rdrdzd\varphi \quad (6)$$

The integration is performed embedding cylinder's volume. The one band Hamiltonian contains only the solutions of the forms T_1 T_2 T_3 and their corresponding Hamiltonian matrix elements can be evaluated as derived in Vukmirovic et al. (2006).

6. APPLICATIONS

QDs Photovoltaic Solar Cells

When photons of sunlight hit a solar cell: They strike electrons in semiconductor material and send them on their way as an electric current. Although many solar photons carry enough energy to theoretically unleash several electrons, they almost never free more than one. An electron loosed by absorbing a photon often collides with a nearby atom. But when it does, it is less likely to set another electron free than it is to create atomic vibrations that squander the electron's excess energy on heat. However, a team of researchers at Los Alamos National Lab found that quantum dots produce as many as three electrons from one high energy photon of sunlight. This could boost the efficiency of solar panels considerably (Weiss, 2006).

Quantum Dot Laser

Since quantum dots absorb and emit light in a very narrow spectral range which can be controlled by an applied magnetic field illustrates potential application in the construction of more efficient and precisely controllable semi-conductor lasers. Arakawa and Sakaki predicted in the early 1980s that quantum dot lasers should exhibit performance that is less temperature-dependent than existing semiconductor lasers, and that will in particular not degrade at elevated temperatures. Other benefits of quantum dot active layers include further reduction in threshold currents and an increase in differential gain, i.e. more efficient laser operation. Stimulated recombination of electron-hole pairs takes place in the GaAs quantum well region, where the confinement of carriers and of the optical mode enhances the interaction between carriers and radiation. Recently Fujitsu and the University of Tokyo have developed a 10 Gbit/s quantum dot laser which is not affected by the temperature changes, and could be used for optical data communications and optical networks and it uses 3-dimensional nano-structured quantum dots in the light-emitting area.

Physicists at the National Institute of Standards and Technology (NIST) and Stanford and Northwestern Universities have built micrometer-sized solid-state lasers in which a single quantum dot can play a dominant role in the device's performance. Correctly tuned, these microlasers switch on at energies in the sub-microwatt range. These highly efficient optical devices could one day produce the ultimate low-power laser for telecommunications, optical computing and optical standards. The team made microdisk lasers bi-layering indium arsenide on top of gallium arsenide. The mismatch between the different-sized atomic lattices forms indium arsenide islands, about 25 nanometers across, that act as quantum dots. The physicists then etched out disks, 1.8 micrometers across and containing about 130 quantum dots, sitting atop gallium arsenide pillars.

The disks are sized to create a whispering gallery effect in which infrared light at about 900 nanometers circulates around the disk's rim. That resonant region contains about 60 quantum dots, and can act as a laser. It can be stimulated by using light at a non-resonant frequency to trigger emission of light. But the quantum dots are not all identical. Variations from one dot to another mean that their emission frequencies are slightly different, and also change slightly with temperature as they expand or contract. At any one time, the researchers report, at most one quantum dot and quite possibly none has its characteristic frequency matching that of the optical resonance.

Nevertheless, as they varied a disk's temperature from less than 10K to 50K, the researchers always observed laser emission, although they needed to supply different amounts of energy to turn it on. At all temperatures, they say, some quantum dots have frequencies close enough to the disk's resonance that laser action will happen. But at certain temperatures, the frequency of a single dot coincided exactly with the disk's resonance, and laser emission then needed only the smallest stimulation. It's not quite a single-dot laser, but it's a case where one quantum dot effectively runs the show (Xie et al., 2007).

Quantum Information Processing

Quantum dots have significant optical applications due to their theoretically high quantum yield and in electronic applications they have been proven to operate like a single-electron transistor and show the Coulomb blockade effect. Quantum dots may also have implementations in quantum information processing, the flow of electrons through the quantum dot can be controlled by applying small voltages and thereby make precise measurements of the spin and other properties therein. With several entangled quantum dots, or qubits, plus a way of performing operations, quantum calculations might be possible.

7. DATA ANALYSIS

Schrödinger's equation (time-independent) is solved for infinite (both 1D and 2D cases) and finite potential well (1D case) systems, where a particle in simple potentials is described for single quantum wells in an ideal system. The effects of changing parameters such as potential, well widths are investigated. All the programs are created in Matlab, due to its speed, excellent graphical capabilities, suitable in-built functions and its ease of use. Through out, effective mass approximation is used (GaAs). All plots contain all the necessary information within the figure such as parameters set for each particular analysis etc., Displacements are expressed in angstrom units and energies in electron volts. Qualitative descriptions are provided for each case. Later, a realistic model of quantum well is solved for single well where the effects of changing parameters such as potential, well widths are investigated and its wave functions are plotted with associated energy levels and their transition between the states. In case for quantum dots, a cylindrical quantum dot is analyzed for fixed radius with changing height, the original Matlab code developed by N Vukmirovic has been modified.

Figure 8 shows the wave functions for a particle in a square box where the wave function is zero at the walls and the solution for the wave function yields just sine waves. There are four energy levels plotted corresponding to their eigen states and energy levels transition plot is right next to the ψ plot. The allowed values of *E* are normally discrete, i.e. energy quantization arises naturally in wave mechanics.

Figure 8. Four numerical solutions of Shrodinger eqn. and their corresponding energy levels

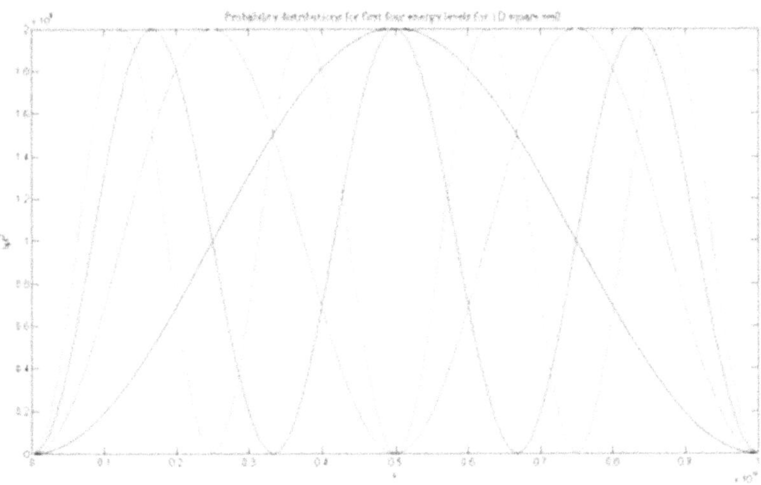

The probability distribution of the wave function squared is plotted above in Figure 8. It can be noticed that the wave function do not cross the region outside the walls and becomes uniformly distributed at higher n. Well width is set at 10 angstrom and calculated for effective mass approximation (me=0.067*m0). Figure 9 is plotted energy of an electron with changing well width. Energy states is inversely proportional to the square of the well width, which implies the energy separation of neighbouring states decreases on increasing well width and gives particle more freedom. As the width approaches infinity, the separation of the neighbouring levels approaches zero and the effect of quantization become negligible. The particle becomes unbounded and free.

Figure 9. Probability distribution of the particle in foreign states for square well

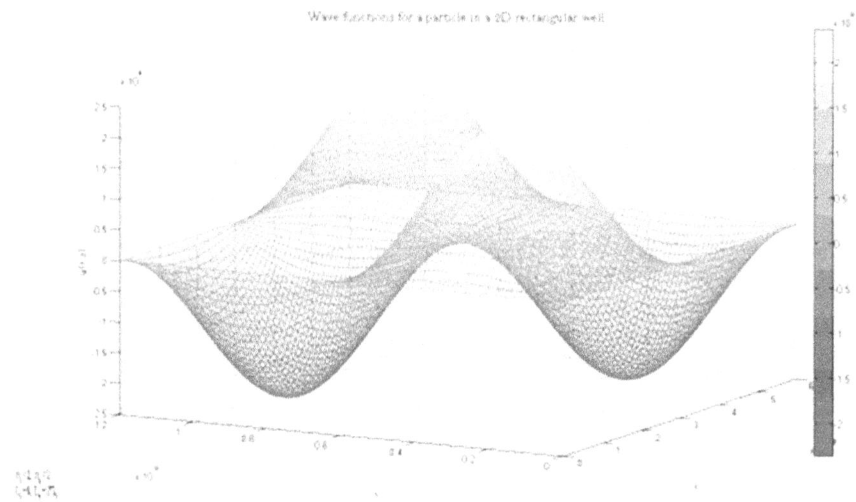

Figure 10. Wave function for a particle in a 2D rectangular well

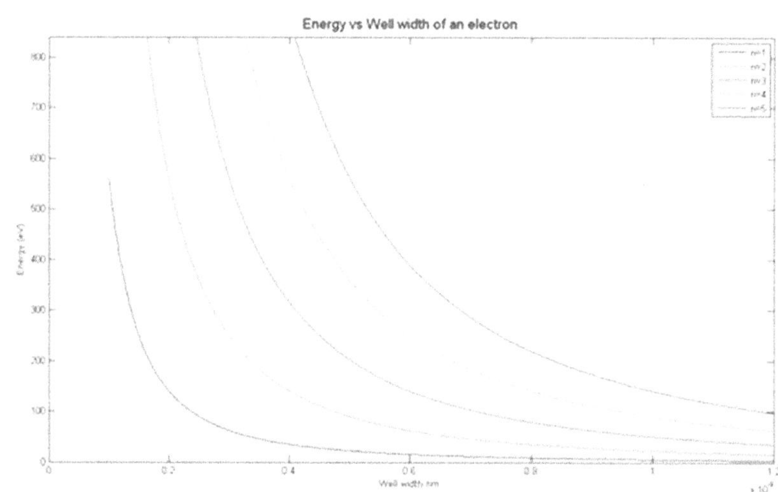

Figure 10 shows the wave function for 2D rectangular well with parameters shown in the fig. The boundary conditions are similar except extended to one extra dimension. The state n1 and n2 are same and are independent of each other.

Figure 11. Wave function for a particle in a 2D rectangular well

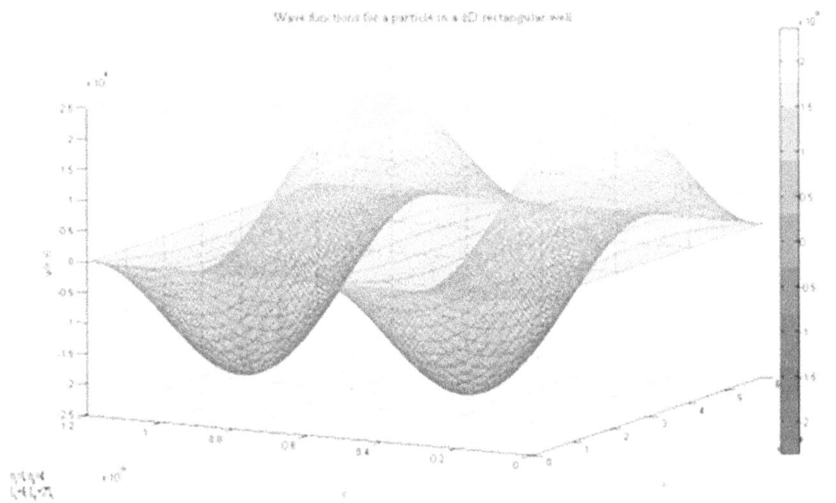

Figure 12. Wave function for a particle in a 2D square well

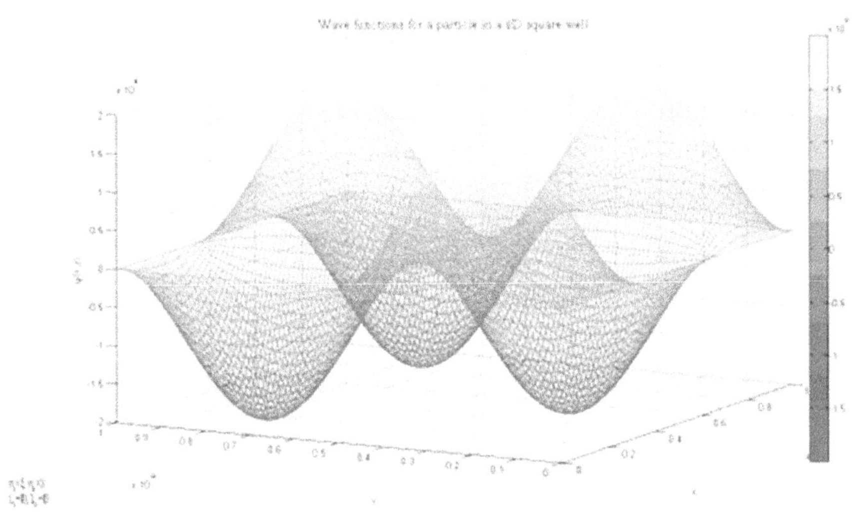

The wavefunction in Figure 12 has different quantum numbers for each state and the rectangular well has different sides ($L_1 = 2L_2$ here). Figure 13 is the wavefunction for a particle in 2D square well with different states but with same sides ($L_1 = L_2$).

Figure 13. Wave function for a particle in a 2D square well

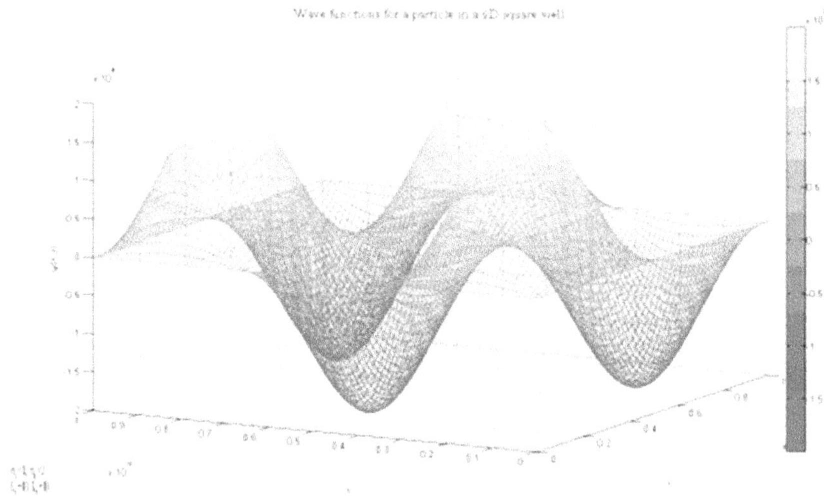

Figure 13 shows the wavefunction with two different states and with same side length. There probability distribution can be plotted as:

Figure 14. Probability distribution shown for the particle in 2D rectangular and square well for the previously plotted functions

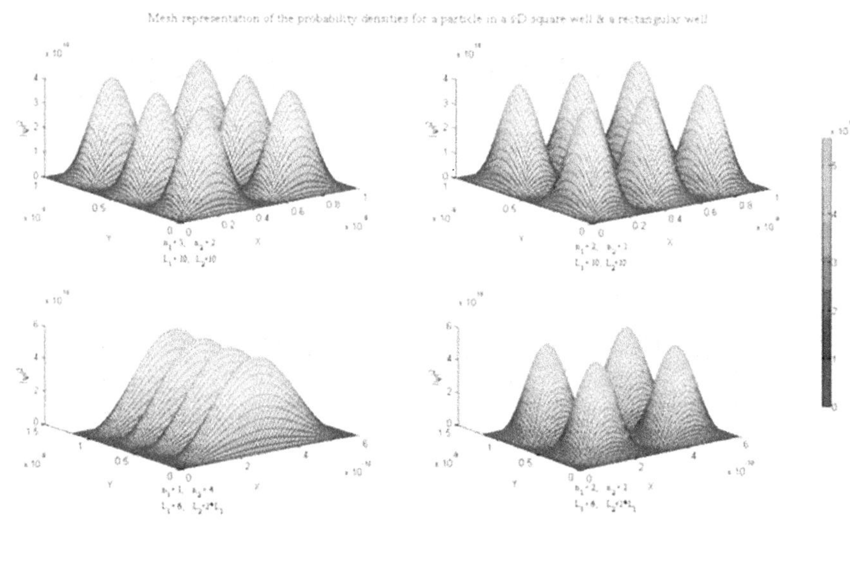

Many features of the 1D system are reproduced in higher dimensions. There is a zero-point energy and the energy separation increases as the walls move apart and become less confining. The energy is quantized as a consequence of the boundary conditions. The contour plot shown in figure 15 shows the degenerate states for two potential wells: Square and rectangular. Degeneracy implies that there is symmetry in the system and are both at the same energy level. An energy level is said to be degenerate if it contains two or more such states. Degenerate eigenstates correspond to identical eigenvalues of the Hamiltonian. If the symmetry is broken by a perturbation, such as applying an external electric field, this can change the energies of the states, causing energy level splitting.

Figure 15. Contour plot representation of the degenerate states in 2D quantum sq. and rect. potential well, mentioned before

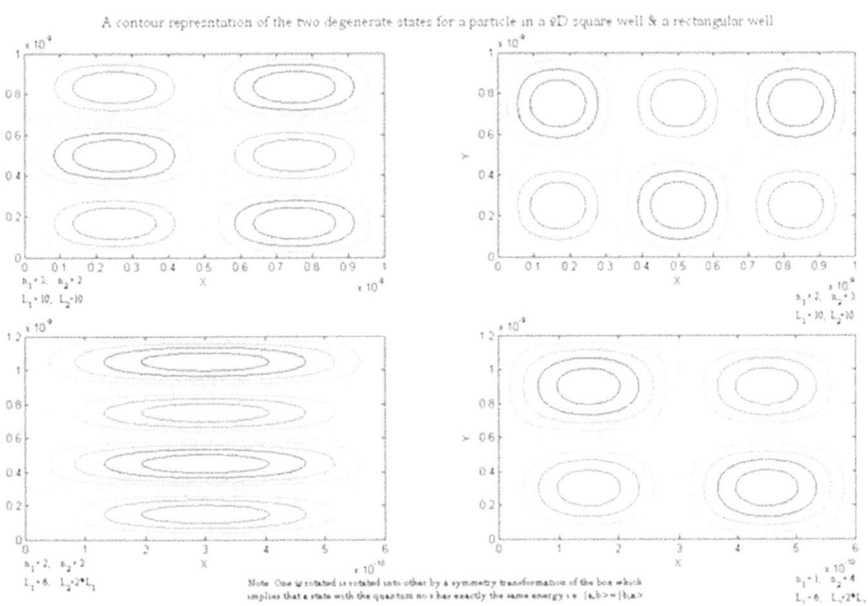

The following points can be implied about bound states for particles:

- The energies are quantized and can be characterized by a quantum number n
- The energy cannot be exactly zero.
- The smaller the confinement, the larger the energy required.

Figure 16. Energy level showing degeneracy of states mentioned for two potential well (square and rectangular well)

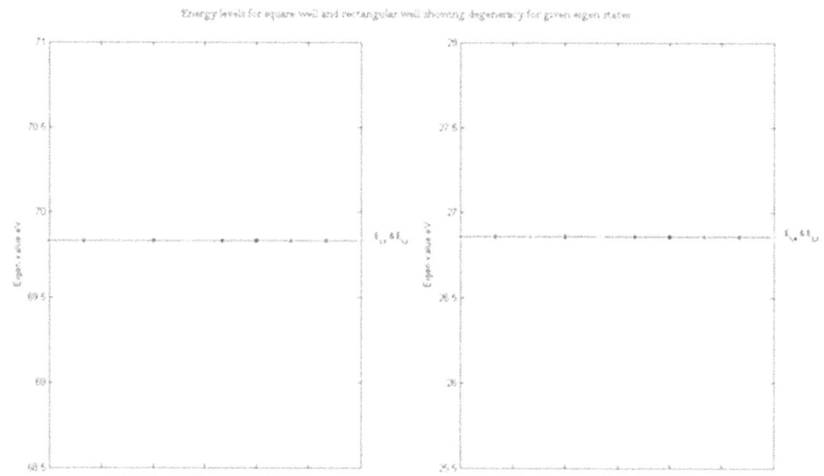

Figure 17. The wave functions of the simple harmonic oscillator for v up to 3

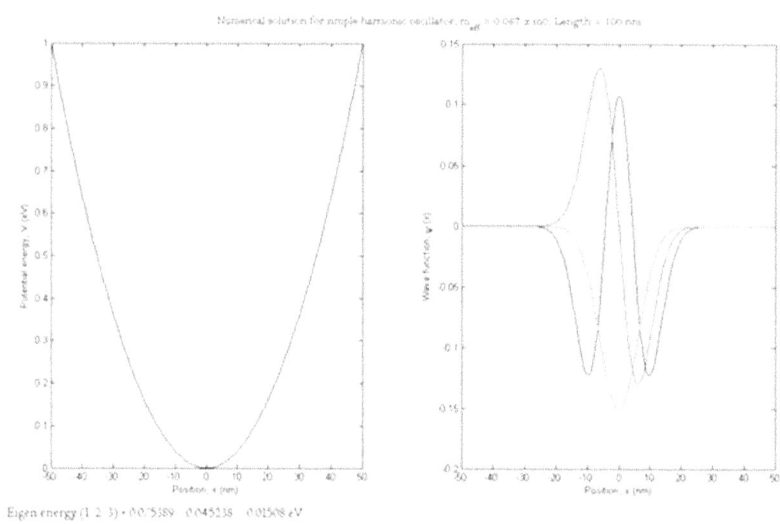

Figure 17 shows the energy levels of the SHO are quantized at equally spaced values. While the energy eigenvalues may be discrete for small values of energy, they usually become continuous at high enough energies because the system can no longer exist as a bound state. For a more realistic harmonic oscillator potential (perhaps representing a diatomic molecule), the energy eigenvalues get closer and closer together as it approaches the dissociation energy. The energy levels after dissociation can take the continuous values associated with free particles.

Figure 18 shows the calculation/finding the bound energies where the potential is changed from 0.3 – 0.4 – 0.5 and then 0.225. F(E) is plotted against these values. Using fzero function in the Matlab, zeros of the function can be found and those values are at which the function is zero represents the bound energies and the number of values per potential represents the number of eigen states present in a given potential well system.

Figure 18. Finding bound energies where the fuction is zero

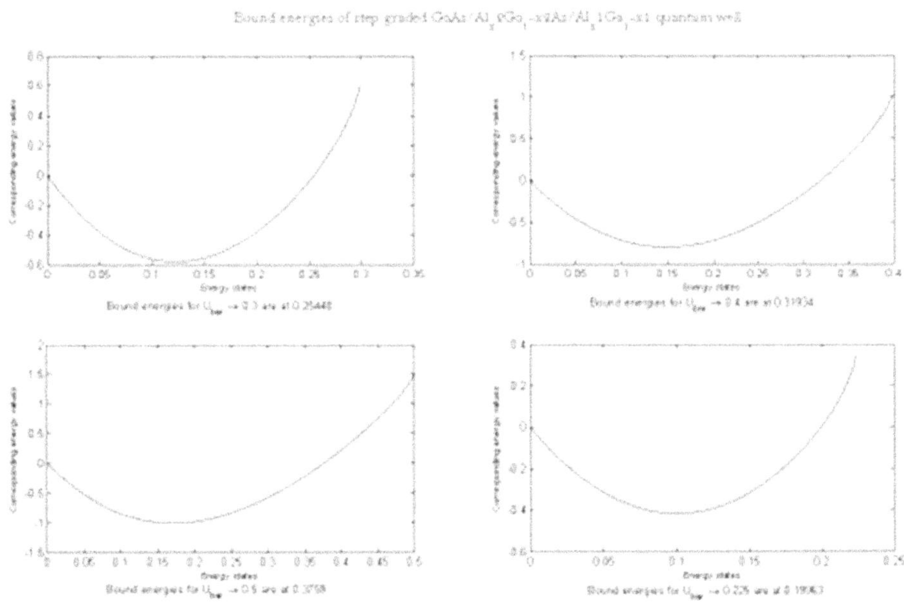

Realistic Quantum Well Model

Shown in the plots above are the bound energies with respect to the change in well width, their corresponding wavefunction and probability densities as well as transition energies w.r.t changing width. The available energies are calculated for the given potential well (0.4, 0.5, 0.225 eV) between the range of well width (10 – 100 Å) and the effect is analysed. Wave functions and their probability for each of the given potential well are plotted for the different states. The model considered is a realistic quantum well. As it can be noticed that the number of states increases on increasing the potential and decreases the energy levels of the states on increasing the well width.

Figure 19. Similar to the one plotted before but the well width changes constantly. The bound energies are at F(E) =(0).

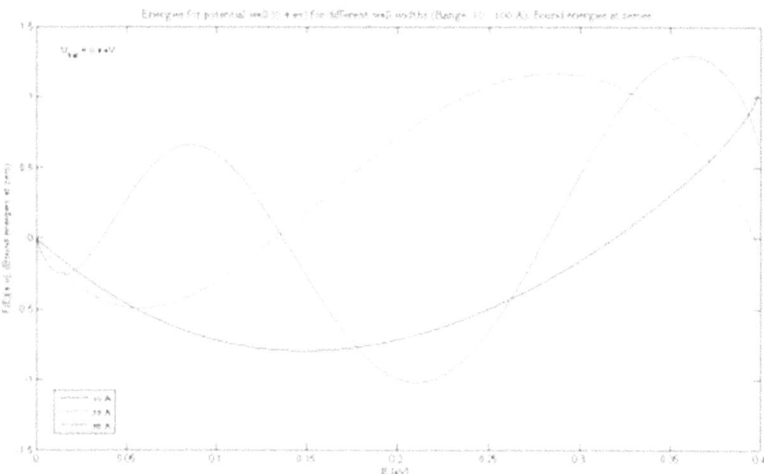

Figure 20. Corresponding wave functions for the bound energies and their probability densities.

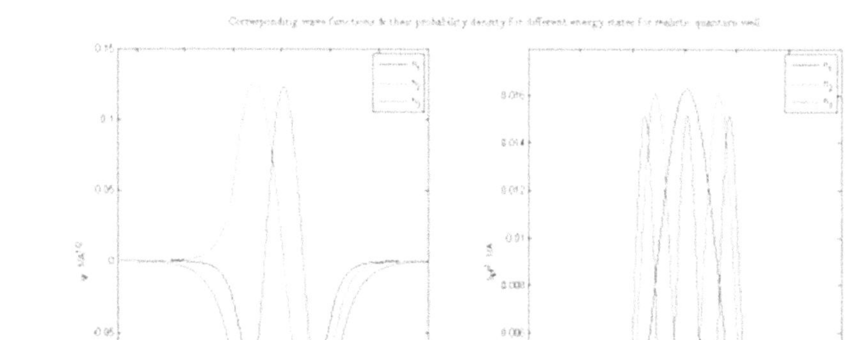

Figure 21. Bound energies w.r.t changing well width and the transition between the energy levels.

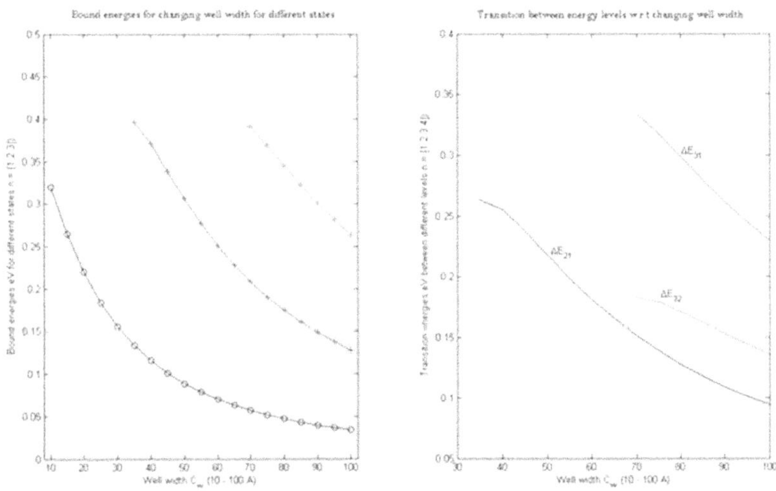

Figure 22. Finding bound energies in a potential well of given Ubar = 0.5 eV.

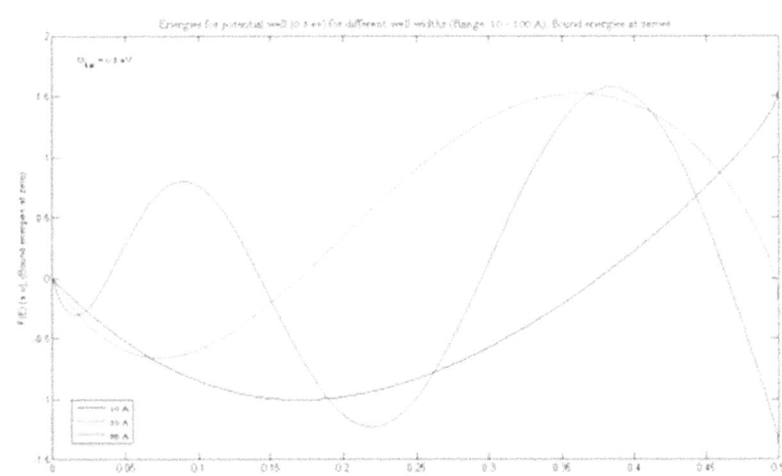

Figure 23. Bound energies w.r.t changing well width and the transition between the energy levels plot.

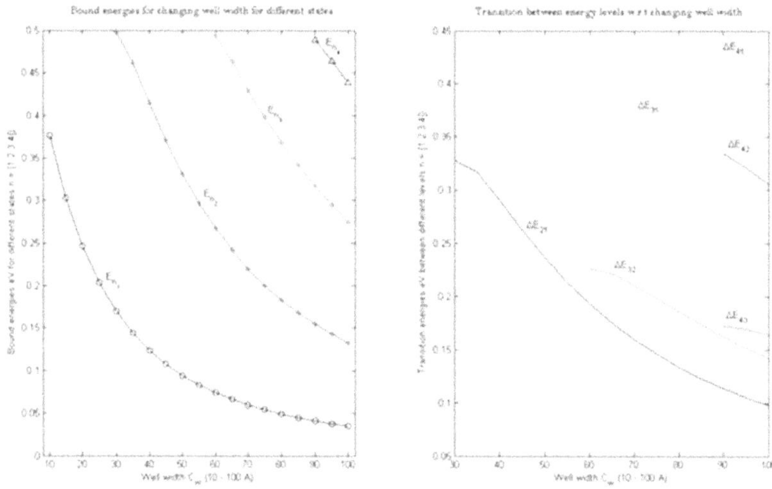

Figure 24. Corresponding wave functions for the bound energies and their probability densities.

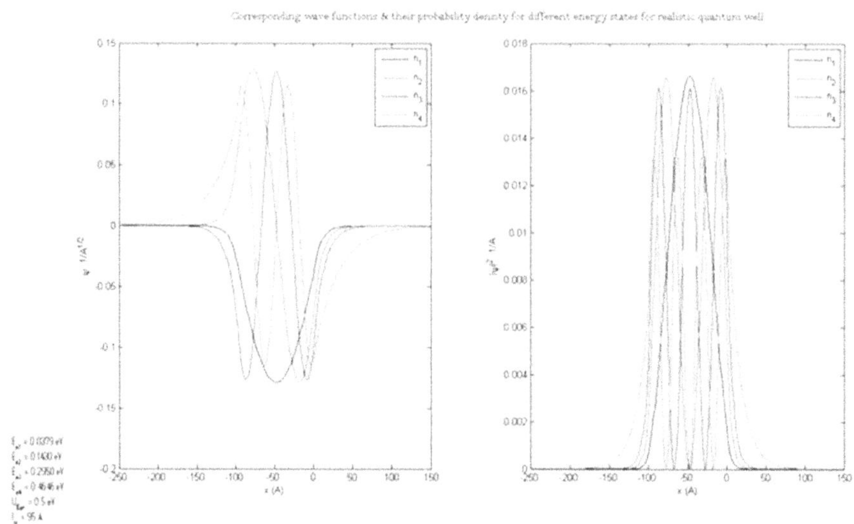

Figure 25. Finding bound energies in a potential well of given Ubar = 0.225 eV.

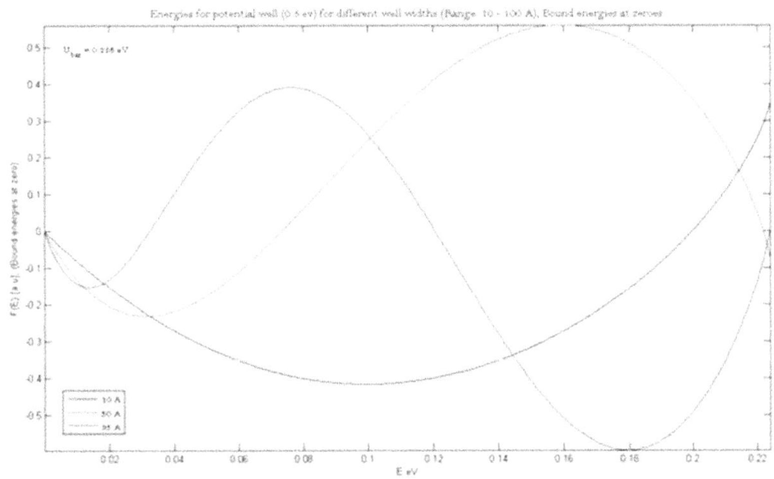

Figure 26. Corresponding wave functions for the bound energies and their probability densities.

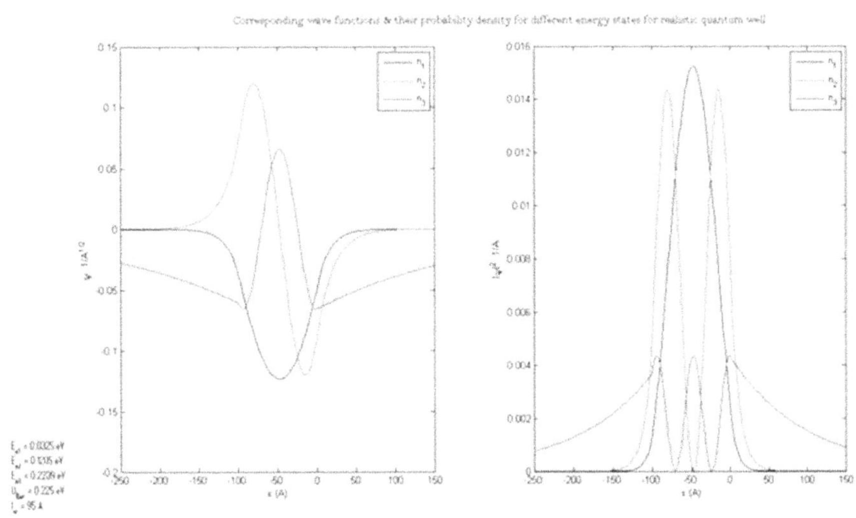

Figure 27. Bound energies w.r.t changing well width and the transition between the energy levels.

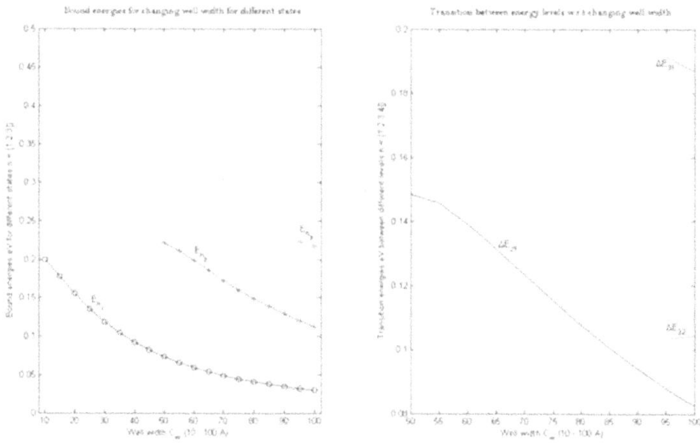

Bound energies per state increases on increasing the potential of the well and also the separation between the levels. Increasing the potential leads to ideal case where the wave function tends to limit within the walls of the well. Penetration depth decreases with the mass of the particle and the height of the barrier (potential) above the energy of the incident particle. Because the particle is confined, its energy is quantized, and the boundary conditions determine which energies are permitted.

The energy state or the quantum number (n) determines the wave functions and the energies. The squares of the wave functions shown represent the probability densities for finding the particle in each state. Note how particle becomes increasingly uniformly distributed on increasing n and the increasing nodes for wave functions that represent higher energies. The energy separation of neighbouring states decreases as the wall width increases and give particle more freedom.

As the length of the box approaches infinity, the separation of the neighbouring levels approaches zero, and the effect of quantization becomes completely negligible. In effect the particle becomes unbounded and free.

Spectroscopy can be used to measure the energy levels of quantum well systems, which involves illuminating the semiconductor with light from laser.

Optical transitions between the states with the discrete energy levels give rise to sharp lines in the absorption and emission spectra. Absorption occurs during the propagation if the frequency of the light is resonant with the transition frequencies of the atoms in the medium. Only unabsorbed light will be transmitted and selective absorption is responsible for the colouration of many optical materials. Luminescence process occurs for the excited atoms which re-emit by spontaneous emission to dissipate the excitation energy and the emitted light has a frequency that depends upon the energy level difference the electrons transition. The optical transitions of the solids are more complicated to deal with. Absorption spectra of free atoms consist of discrete lines but in solids where the atoms are packed very close to each other overlapping the outer orbitals, makes interaction strong with each other and the discrete levels of free atoms broadens into bands where the optical transition occur between the electronic bands. The wave functions for the solids are described by Bloch's wave function that has periodicity of the lattice. For direct band gaps semi-conductors, luminescence strongly when the electrons are promoted to the conduction band and the luminescence wavelength coincides with the band gap of the semi-conductor, which forms the physical basis for the light-emitting devices.

Cylindrical QD

Figure 28. Wave function modulus for the cylindrical quantum dot where r = 80 A and across Z shows the probability density of QD

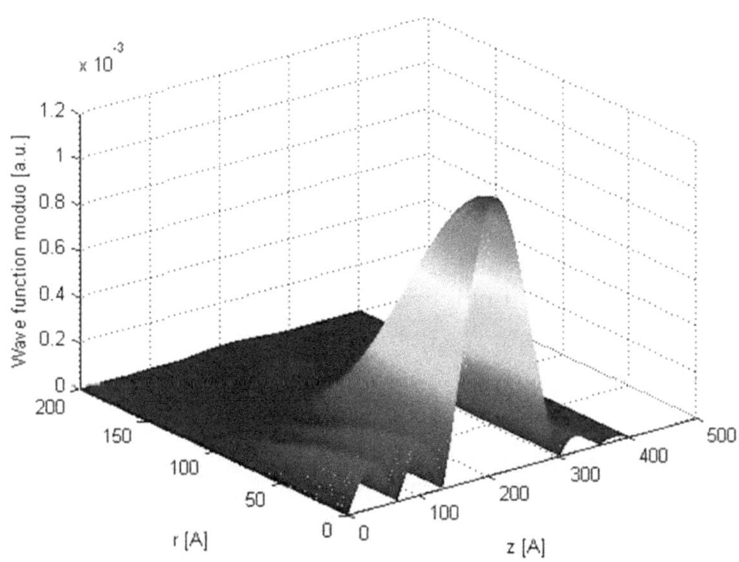

Figure 29. Bound energies and transition between the energy level plot for cylindrical quantum dot where r = 80 A

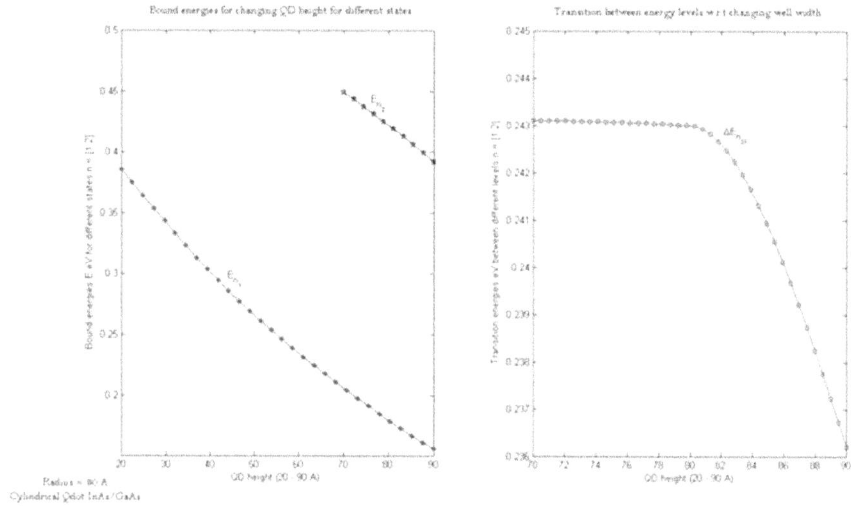

8. CONCLUSION

The Schrodinger equation is an extremely powerful mathematical tool and the whole basis of wave mechanics. It gives the quantized energies of the system and the form of the wave function so that other properties may be calculated. Quantum well heterostructures can increase the strength of electro-optical interactions by confining the carriers to small regions. For the finite potential well, the solution to the Schrodinger equation gives a wave function with an exponentially decaying penetration into the classically forbidden region. Also as the value of n increases i.e. at higher state the particle is uniformly distributed. Confining a particle to a smaller space requires larger confinement energy. Since the wave function penetration effectively enlarges the box, the finite well energy levels are lower than those for the infinite well.

The energy separation of neighboring states decreases as the walls move apart and give particle more freedom. As the length of the box approaches infinity, the separation of neighboring levels approaches zero, and the effect of quantization become completely negligible and the particle becomes unbounded and free. In other case smaller the confinement, the larger the energy. The penetration depth decreases with the height of the mass of a particle and the height of the barrier above the energy of the incident particle. With the infinite well, one can increase energy indefinitely and always encounter more bound states. With the finite well, however, the particle is no longer confined when energy reaches the potential well depth and there are no more bound states. The number of bound states depends on the well depth and width, but it is always finite.

In harmonic oscillator, there is symmetrical occurrence of momentum and displacement in the expression for its total energy. The significant point about the energy levels is that they form a ladder with equal spacing. The energy separation between neighbors is regardless of the value of υ. As the force constant influences the separation between neighboring levels decreases, greater the force constant, the more confining the potential and therefore the more sharply curved the wave functions.

The electron energy levels in quantum dot are discrete, the addition or removal of atoms to the quantum dot has the effect of altering the boundaries of the bandgap. Altering the geometry of the surface of the quantum dot also changes the bandgap energy, owing again to the small size of the dot, and the effects of quantum confinement.

The coloration is directly related to the energy levels of the quantum dot. The bandgap energy that determines the energy (and hence the color) of the fluoresced light is inversely proportional to the square of the size of the quantum dot. In general, quantum dots can be engineered for specific electronic and optical properties by controlling the geometrical size, shape, and the strength of the confinement potential.

Through out particles are treated that move non - relativistically in 1D where the actual case for all real systems are in three-dimensions, even so the main significance of one-dimensional systems is that they provide a good introduction to three-dimensional systems, and that several one-dimensional solutions find direct application in three dimensional problems.

REFERENCES

Atkins, P. W. & Friedman, R. S. (2010). *Molecular Quantum Mechanics.* Oxford University Press.

Bryant, G.W, & Solomon, G.S. (2004). *Optics of Quantum dots and wires.* Artech House, Inc.

Cleve B. Moler, C. (2004). *Numerical Computing with Matlab.* Siam Publishing.

Daku, B. (2006). *Matlab tutor CD.* John Wiley & Sons.

Fox, M. (2010). *Optical properties of Solids*. Oxford University Press.

Hanselman, D & Littlefield, B. (2000). *Mastering Matlab 6 – A Comprehensive tutorial and reference*. Prentice Hall.

Harrison, P. (2001). *Computational methods in Physics, Chemistry & Biology: An Introduction*. John Wiley & Sons.

Harrison, P. (2016). *Quantum Wells, Wires and Dots: Theoretical and Computational Physics*. John Wiley & Sons.

Hinchliffe, A. (1988). *Computational Quantum Chemistry*. John Wiley & Sons.

Jacak, L. & Wojs. H. (1998). *Quantum dots*. Springer.

Levi, A. (2012). *Applied Quantum Mechanics* (2nd ed.). Cambridge University Press.

Liboff, R. (2002). *Introductory Quantum mechanics (4th edn.)*. Pearson education Inc.

Michler, P. (2003). *Single quantum dots: Fundamentals, applications and new concepts*. Springer.

Pratap, R. (2009). *Getting started with Matlab – A quick introduction for Scientists and Engineers*. Oxford University Press.

Thijssen. J. M. (2007). *Computational Physics* (2nd ed.). Cambridge University Press.

Van Loan, C. (1996) *Introduction to scientific computation – A Matrix vector approach using Matlab*. Prentice Hall.

Vukmirovic, N., Gacevic, Z., Ikonic, Z., Injin, D., Harrison, P., & Milanovic, V. (2006, July). Intraband absorption in InAs/GaAs quantum dot infrared photodetectors –effective mass versus k x p modelling. *Semiconductor Science and Technology*, 3.

Weiss, P. (2006). Quantum-Dot Leap. *Science News Week, 169*(22).

Wong, S. (1997). *Computational methods in Physics & Engineering* (2nd ed.). World Scientific.

Xie, Z. G., Götzinger, S., Fang, W., Cao, H., & Solomon, G. S. (2007). Influence of a single quantum dot state on the characteristics of a microdisk laser. *Physical Review Letters, 98*(11), 117401. doi:10.1103/PhysRevLett.98.117401 PMID:17501091

Chapter 6
An Analysis of Quantum Computing Spanning IoT and Image Processing

P. Kamaleswari
SRM Institute of Science and Technology, India

A. Daniel
https://orcid.org/0000-0001-5564-2332
Amity University, India

ABSTRACT

The significance of using quantum computing in a variety of applications is illustrated by qubits, coherence time, and gate error rate, respectively, in all of the measurements. In quantum information, the fundamental units are called qubits, which are comparable to bits in classical information. The complicated computation can be simplified by employing qubits, which can be used in a wide variety of contexts. Data can be analyzed without producing any decoherence errors, which is particularly useful for image processing and real-time applications. The concept of "noisy intermediate scale quantum" (NISA) can be addressed and used in real-time data collecting through the Internet of Things as well as in picture processing. The NISA has a primary emphasis on factorization and optimization techniques for the purpose of data analysis. Therefore, this progress will be essential in unlocking the full potential of quantum computing and enabling it to tackle a wider variety of difficult challenges like the Internet of Things (IoT) and Image Processing

DOI: 10.4018/978-1-6684-7535-5.ch006

Copyright © 2023, IGI Global. Copying or distributing in print or electronic forms without written permission of IGI Global is prohibited.

1. INTRODUCTION

Even NP-Complete problems of exponential complexity can be solved by quantum computers at a rate we previously thought to be inconceivable. Quantum algorithms make it possible to address issues that are too challenging for classical computers to handle (e.g., huge number factorization, brute-force search). In both academia and industry, quantum technologies are expanding quickly (sometimes known as the "race to the quantum computer"). For instance, organizations and research centers around the world have developed quantum hardware for a key distribution network. The ongoing advances in quantum computing are the result of interdisciplinary research. Quantum computing has spread equally in various applications across multiple fields by solving complex problems more efficiently than classical computers. Generally, large-scale quantum computers are still in development, researchers and industry experts are exploring several areas where quantum computing could make a significant impact. The discussing of quantum computing could be introduced in different applications. The performance of a quantum computer is typically measured by its number of qubits, The more qubits a quantum computer has, the more complex computations it can potentially handle. Coherence time refers to the duration during which a qubit can maintain its quantum state without significant errors or decoherence. Coherence times that are longer make it possible to do more intricate computations. The efficiency of quantum operations on qubits is quantified by the gate error rate. If you want your computations to be precise and dependable, you should aim for low gate error rates. With just the API, the FCQ-CNN model achieved an accuracy of (Acc-84.6%) when tested on IBM's state-vector simulator. However, transmission across large qubits is only possible for a limited amount of data samples relative to the total number of qubits (Simões et al., 2023). QNN plays improved upon previous efforts by elevating QSVMs by up to 5%.Analysis shows that the QNN outperforms the QSVM by a margin of 5%, proving that quantum implementations are superior to hybrid quantum-classical ones (Islam et al., 2022). Algorithms based on quantum intelligence have shown to be very competitive and boast impressive processing times. This method effectively sidesteps the limitations of intelligent algorithms by combining the strengths of global search with the benefits of high parallelism, powerful storage, and computation offered by quantum computing. The number of clustering categories is not assumed and instead is determined only by the data's potential information (Ren et al., 2020). The focus of the paper is on identifying patterns in the data and creating clusters out of sets of related observations. In this case in particular, the quantum-inspired k-means algorithm outperforms the Qk-means algorithm by a margin of +3 to +5% in terms of accuracy. When compared to the top classical algorithms, the complexity of the Bernstein-Vazirani (BV) algorithm scales better (Khan & Robles-Kelly, 2020). In particular, compared to the best classical method, the number of requests processed by the quantum algorithm will increase linearly. The depth and noise of a quantum circuit may have an effect on the model's accuracy and training duration across industries, and this work can be generalized to encompass other quantum machine learning algorithms (Raleganka et al., 2021). The suggested system's QML-based intrusion detection algorithm's training time has been cut by more than half, improving the system's efficiency. Quickly learning QML models will require future work on optimizing quantum algorithms and parallelization strategies (Singh et al., 2021). When compared to a quantum NN and an LSTM NN, a hybrid quantum-classical NN is superior for detecting cyberattacks within a vehicle. LSTM NN achieves 99.96% accuracy on the training dataset, while HQCNN achieves just 98.73% accuracy. In this way, more time can be spent maximizing the efficiency of dependable production (Li et al., 2020). In a classification protocol called optimal QKD, the RF classifier demonstrates its remarkable advantages in accuracy, resilience, and efficiency, leading to an overall success rate of 98.2%. In the near future, QKD will be used in real-time in widespread applications of the multi-user QKD-network.

1.1 Cryptography

Quantum computers have the potential to break many of the encryption algorithms used to secure sensitive information today. However, they also offer new cryptographic techniques that are resistant to quantum attacks. Quantum key distribution (QKD) enables secure communication by using the principles of quantum mechanics to exchange encryption keys (Raleganka et al., 2021). Quantum-resistant algorithms are also being developed to protect data from potential quantum threats.

1.2 Drug Discovery and Material Science

Quantum computers can simulate and analyze complex molecular interactions, allowing for more efficient drug discovery and material design. Quantum algorithms could help in simulating molecular structures and properties, enabling researchers to identify promising drug candidates or optimize materials with specific properties, such as improved battery performance or stronger materials.

1.3 Optimization and Logistics

Quantum computing can provide significant advantages in solving optimization problems. For example, it could be used to optimize supply chain management, transportation routes, or scheduling in various industries. Quantum algorithms, such as the Quantum Approximate informed investment decisions. Optimization Algorithm (QAOA) and Quantum Annealing can potentially provide faster and more accurate solutions to these optimization problems.

1.4 Financial Modeling and Portfolio Optimization

Quantum computing could be applied to complex financial modeling, risk analysis, and portfolio optimization. It could enable more sophisticated analysis of market trends, portfolio diversification, and risk assessment, potentially leading to more informed investment decisions.

Figure 1. A computational diagram showing how quantum computing developed from classical computing. The outcomes of quantum computing are utilized by quantum computers.

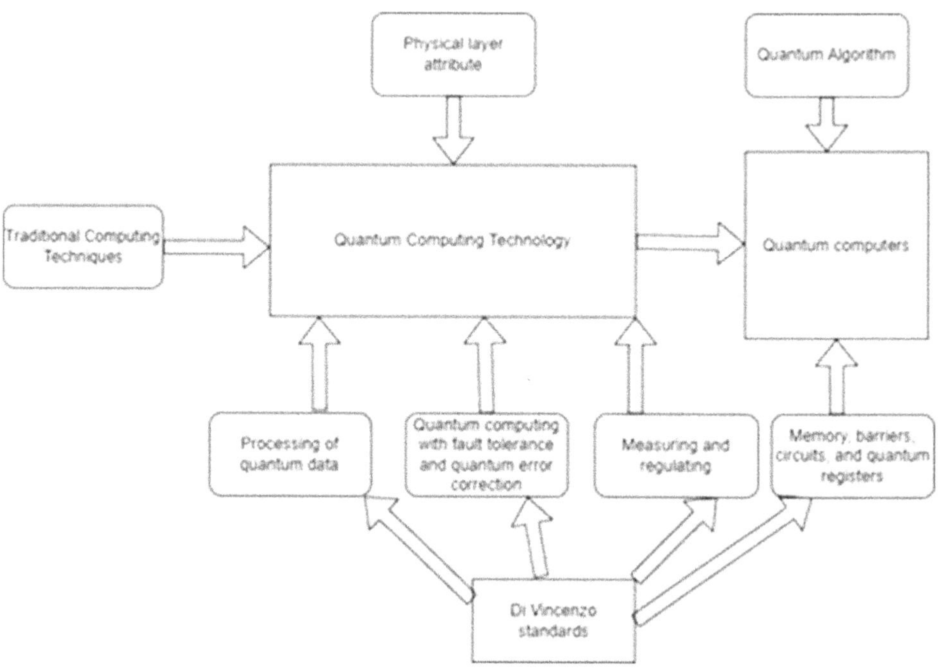

Figure 1 explains about Vincenzo standards contains processing of quantum data, quantum computing with fault tolerance and quantum error correction, measuring and regulating in memory, barriers, circuits and quantum registers.

1.4 Machine Learning and AI

Quantum computing has the potential to enhance machine learning and AI algorithms by providing faster data processing and more efficient pattern recognition. Quantum machine learning algorithms, such as quantum support vector machines or quantum neural networks, could help improve tasks like data classification, clustering, and pattern recognition.

1.5 Energy and Climate Modeling

Quantum computing could contribute to more accurate and detailed modeling of energy systems, climate change, and environmental impact assessments. It could assist in simulating complex quantum mechanical processes involved in energy production, carbon capture, and climate modeling, allowing for better optimization and policy planning.

2 QUANTUM IoT

Numerous cryptographic algorithms now employed to protect data transit and storage may be broken by quantum computers. The importance of protecting IoT devices from quantum attacks rises as they grow more common and linked. In order to develop post-quantum cryptosystems, and new cryptographic algorithms that can fend off attacks from quantum computers, academics are working hard in this area.

The Database collects all the real time data through different types of sensors. From the database collection, the processing of data can be performed by quantum computing means the output prediction can be analyzed with good accuracy and fast retrieval manner.

2.1 QIoT in Agriculture

Quantum computing through IOT has the potential to transform agriculture is enormous. Different types of sensors can be fixed in soil, plants to identify the suitable climate for precision farming, soil moisture, Food chain supply chain management and plant breeding.

Plant breeding: Quantum computing through IOT can help in fast processing of genome sequencing and analysis, which is crucial in identifying desirable traits in crops. This could help breed more disease-resistant, drought-tolerant, and high-yielding crops.

Precision farming: Quantum computing can help farmers make better decisions about Through IOT Quantum computing can help farmers to identify suitable fertilizers for healthy plant growing which yields in sufficient amount of grains and identify the perfect time for plant harvasting. The above processing analyzing can be done through soil moisture sensors, plant disease identifiers and drought sensors etc.

Figure 2. Different kinds of agricultural sensors along with quality control Quantum computing and the various types of sensors used in the agricultural field

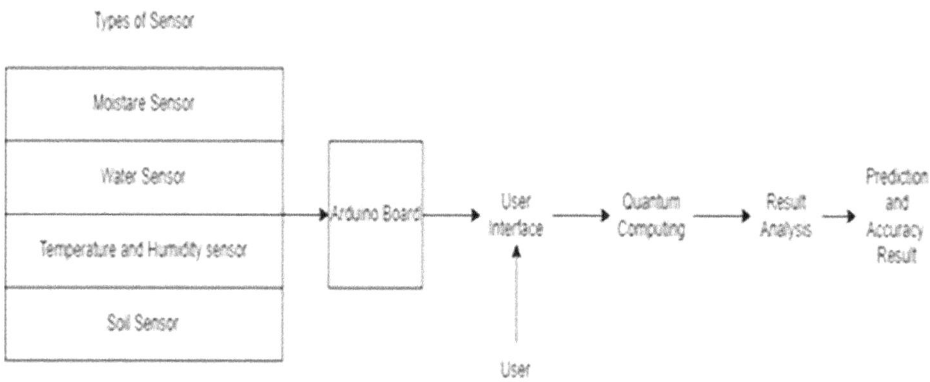

In Figure 2 crop prediction, suitable climate condition for particular crop prediction, flood detection, drought prediction, etc. are the important application performing through quantum computing with the help of sensor values.

Food Supply Chain Management

The healthy cultivated crop can be supply all over area without any deviation in market price with the help of IOT sensor. Once the price of the product is fixed means it cant be any deviation in price based upon the price checklist. Through this application can able to get healthy product to all over world.

2.2 QIoT in Healthcare

IoT activates major role in healthcare modern revolution. Here sensors can be fixed all over the patients and collected real time data can be analyzed whether the particular patient in normal or abnormal condition with the help of quantum computing

Figure 3. Abnormal prediction through IoT sensor

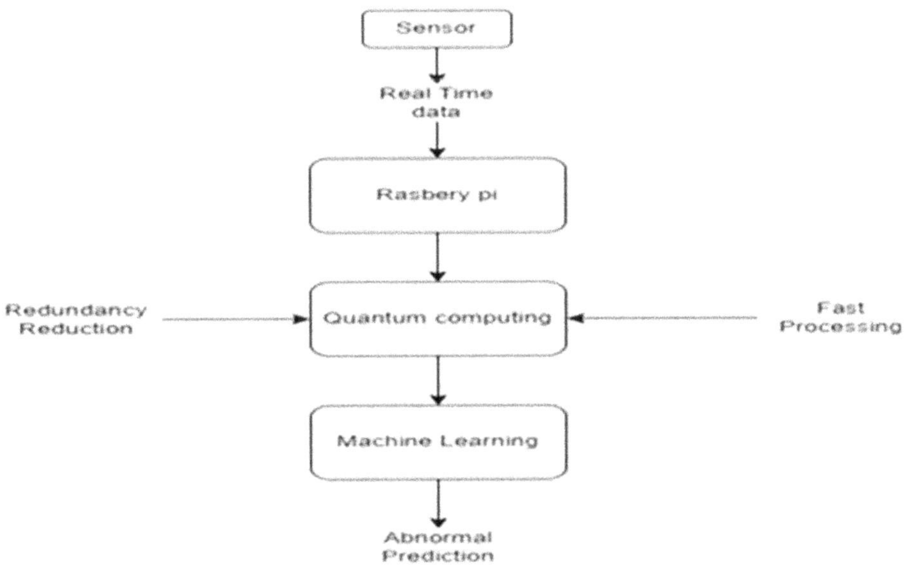

Explains in detail how quantum computing can be used to forecast whether or not a person would exhibit abnormal behaviour. The Raspberry Pi can collect data from all of the sensors, process it using quantum computing, and display the results.

QIoT in Critical Care

If the sensor in critical care patient, here also the real time collected data can be analyzed under quantum computing which can be processed for indicate emergency condition. So immediately the patient attending doctor can aware of it through IOT and attention will be provide to overcome from the critical condition.

Figure 4. QIoT in emergency care patients

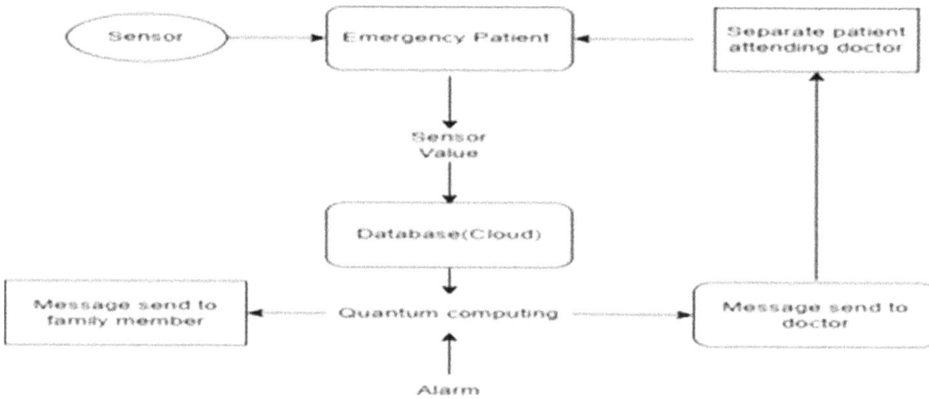

Patients in emergency treatment equipped with sensors using the Quantum Internet of Things.

2.3 QIoT With Image Processing

Through sensor, image of different person, particularly face can be identified and processed by quantum computing with the help of Face recognition algorithm. Quantum computing can process the face image quickly to indicate whether the particular person can authorized person or not. So the external person, who is not belongs to army person database can identified easily and strict authorization can be provided with the help of quantum computing.

Identifying and authenticating people by their facial traits is the primary goal of face recognition, a subfield of image recognition. QIoT is being incorporated into facial recognition systems.

3. QUANTUM IMAGE PROCESSING

Quantum computing has become a cutting-edge field in the era of exponential technological developments. Quantum image processing is an attractive field of research, despite the fact that its potential influence spans several industries. Bringing the potential of quantum computing to the visual domain brings up previously unimaginable possibilities, from improved picture processing to cutting-edge compression methods. In this essay, we explore the interesting field of quantum image processing, as well as its fundamental ideas and prospective applications.

Figure 5. Image recognition through QıoT

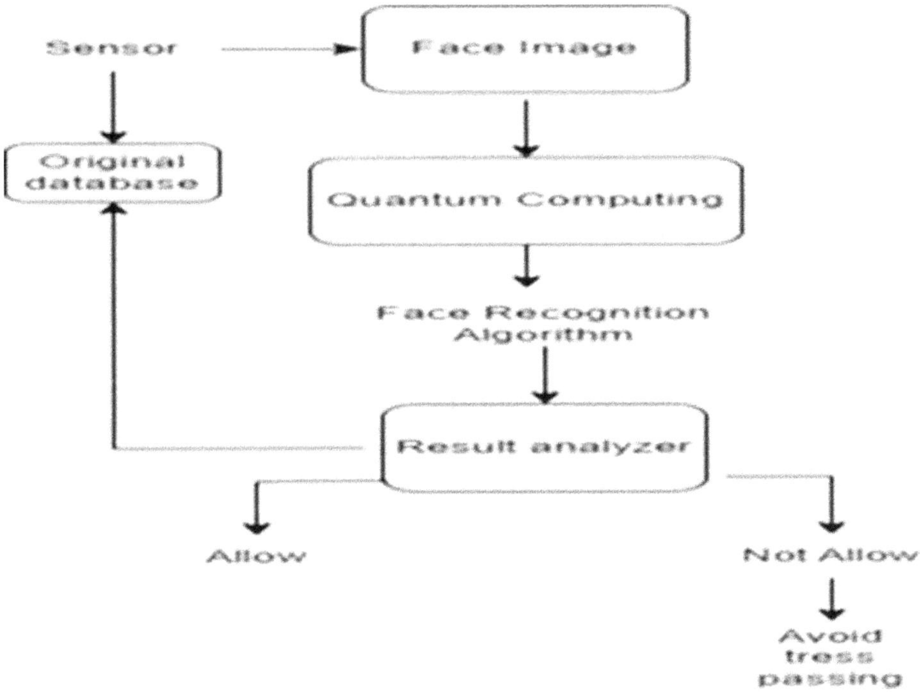

3.1 Qubit Lattice

The idea of a qubit lattice an array of qubits organized in a grid-like structure is one exciting direction within quantum image processing. A qubit lattice has the potential to revolutionize picture analysis, manipulation, and comprehension by utilizing the special qualities of qubits and their entanglement (Ranjan et al., 2020). The core components of quantum computing, known as qubits, have the amazing capacity to exist in superposition states, which concurrently represent multiple values.

3.2 Quantum Superposition and Entanglement

Superposition and entanglement are fundamental concepts in quantum image processing. Multiple features of a picture may be simultaneously represented and altered thanks to quantum superposition. Quantum algorithms provide a huge advantage in terms of speed and efficiency since they can process all possible pixel combinations concurrently as opposed to processing each individual pixel one at a time.

Quantum entanglement also makes it possible to correlate pixels such that their states interact. Enhancing picture recognition and pattern matching is possible because to this characteristic (Khan & Robles-Kelly, 2020). Quantum image processing algorithms take advantage of entanglement to extract complex characteristics and connections between picture components that would be difficult to achieve with traditional image processing methods.

3.3 Flexible Representation for Quantum Images

The colors and accompanying pixel coordinates of the image are recorded using this technique in a normalized quantum state. By using a tensor product, the normalized state is subsequently included in the quantum state. An effective FRQI implementation is known to exist and only requires a computational cost of 24n for n*n images, according to the polynomial preparation theorem (PPT), one of the key results in QIP.

Table 1. Computable complexities

Processes	Computable Complexity
FRIQI	$O(2^{4n})$
NEQR	$O(2^{n})$
Classical	O(n)

Quantum Algorithms for Image Processing

The development of quantum algorithms designed particularly for image-processing applications is ongoing. These algorithms execute tasks like image compression, denoising, and feature extraction more effectively than conventional methods because they make use of quantum computing properties.

A common approach utilized in many quantum image processing techniques is the quantum Fourier transform (QFT). In order to help with tasks like picture enhancing and compression, QFT can effectively analyze the frequency content of an image. Another useful tool is the quantum phase estimation technique, which enables more exact estimates of pixel phases and more accurate picture reconstruction.

3.1. QML

Data optimization, pattern identification, and analysis have the potential to be revolutionized by QML.

The two main components of machine learning difficulties are handling enormous volumes of data effectively and creating algorithms that analyze this data as quickly as feasible. Quantum registers are far more data-efficient than conventional registers for the first job. While a classical bit register of size n can only store binary strings of size n, a qubit register of size n can store two times as many strings of size n by encoding them in amplitudes.

Incorporating machine learning into quantum computation can be done in four distinct ways. The purely classical strategy, the Quantum-Classical strategy, the Quantum-Quantum strategy, and the Classical-Quantum strategy.

3.2. QFT

To help with activities like picture enhancement and compression, QFT (Quantum Fourier transform) can effectively analyze the frequency content of an image. Another useful tool is provided by quantum phase estimation techniques, which precisely estimate pixel phases and enable more accurate picture reconstruction.

3.3. Quantum Object Detection

Grover's method and other quantum search-based image recognition techniques have shown promise for quick and accurate object recognition in photos. The field of object detection is typically broken up into three subfields. The tracking stage of human activity monitoring in video surveillance involves a number of steps, including item detection in any given image or video frame, feature extraction from the detection zone, and classifier training for object classification in the necessary class (Kamaleswari & Krishnaraj, 2023).

3.3.1 Quantum-Inspired Classical Algorithms

Although large-scale quantum computers are still in their early phases of development, researchers have created classical algorithms that are inspired by quantum theory. In order to gain computational benefits inside traditional computing frameworks, these algorithms aim to mimic the behavior of quantum systems. Classical picture compression methods with quantum-inspired designs have shown promise for higher compression ratios and lower data loss.

Figure 6. Extraction, analysis, and application of information from images received from various detection systems constitute computational image processing

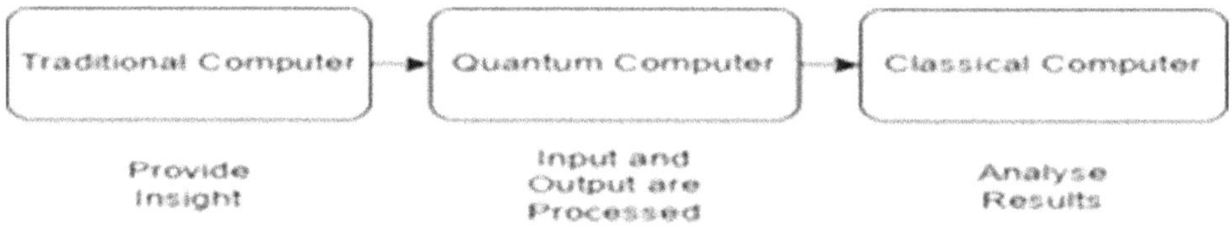

A fascinating nexus between quantum computers and the visual world is quantum image processing (Sachdeva et al., 2023). Quantum algorithms and classical algorithms inspired by quantum theory have the potential to revolutionize picture analysis, compression, and recognition by taking advantage of superposition and entanglement. The prospects and consequences for the future of quantum image processing are limitless, even though there are still obstacles to overcome on the road to practical implementations.

4. NEED FOR THE STUDY AND APPLICATION

It is necessary for research to enhance the image from the original to the enhanced image in order to simplify image analysis and knowledge of visual interpretation. Studying Quantum IoT is essential to overcoming the constraints of conventional IoT systems and utilizing quantum technology's potential for greater security, efficiency, scalability, sophisticated sensing, and future-proofing.

The enhancement technology simplifies the interpretation of the original image and improves how it is seen. There are various obvious approaches to improving technology, but they need a great deal of patience and careful examination. It is also possible to describe enhancement technology as an art form that highlights application-based qualities and helps images to become more resilient. Statistical features can be used to emphasize the characteristics of digital photographs through histogram analysis of augmented images. Researchers have created a number of tried-and-true methods to improve the image simply by expanding the digital image during the last few years. Use interpolation to remove the effects of aliasing on the image. Consequently, it is essential to create an efficient methodology using conventional improvement techniques.

4.1 QIoT Application

4.1.1 Secure Communication

Providing safe channels for data exchange amongst IoT gadgets is a major use case for QIoT. Unbreakable encryption and safe key exchange are possible thanks to quantum technologies like quantum key distribution (QKD).

4.1.2 Sensing and Monitoring

When applied to IoT systems, QIoT can improve their sensing and monitoring capabilities. The use of quantum-enhanced measurement techniques in quantum sensors allows for more precise and accurate readings of physical quantities. This allows for more precise and timely analysis of data during monitoring in fields as diverse as environmental science, smart agriculture, healthcare, and industrial sensing.

4.1.3 Quantum-Enabled Edge Computing

Quantum Internet of Things (QIoT) allows for the use of quantum computing at the network's periphery. Edge devices with quantum computing capabilities may do in-depth data analysis and computations locally, drastically cutting down on network latency and bandwidth needs. This can allow for instantaneous processing and decision-making in time-sensitive Internet of Things applications.

4.2 QIP Application

4.2.1 Recognizing Edges

Quantum image processing's earliest uses were in areas like Edge Detection. Quantum versions of image processing software are required for edge detection.

Fourier transforms, Hadamard transforms, and Haar wavelet transforms are used as examples of fundamental classical picture transformations. In quantum settings, these transformations have polynomial complexity (O(n)), but in classical settings, they have exponential complexity (O(n2)).

4.2.2 Segregating Images

The goal of quantum image segmentation is to improve the accuracy and efficiency of picture segmentation by taking use of the quantum nature of quantum systems. Potential uses of quantum computers for picture segmentation include the following:

- Quantum-inspired Clustering
- Quantum-based Feature Extraction
- Quantum Graph Cut Methods

Explore quantum-inspired image segmentation machine learning models. These models interpret image data and build segmentation masks using quantum circuit topologies, gates, and variational algorithms. These models may improve image segmentation by using quantum computers' parallelism and optimization capabilities.

4.2.3 Image Compression

Multimedia storage and transmission, for example, rely heavily on effective image compression. In order to compress images more efficiently without sacrificing quality, quantum image compression methods make use of quantum computing.

4.2.4 Image Enhancement

Improved image quality and finer levels of detail are possible with quantum image processing. Denoising, sharpening, and contrast enhancement are just a few of the tasks that may be accomplished by quantum algorithms thanks to their use of qubits and quantum gates. This can be especially helpful in the field of medical imaging, as improved image quality can aid in the diagnosis and planning of patient care.

4.2.5 Quantum Watermarking

Digital photos can have their copyright and authenticity safeguarded via watermarking. Since quantum principles are used in both the watermarking process and its extraction, quantum watermarking systems are more secure and resistant to attacks. Quantum watermarking allows for the unbreakable attribution of ownership and identity of images.

5. SUMMARY AND PROSPECTS

This paper provided a high-level introduction to quantum computation, its applications in image processing and Internet of Things research, and the ways in which it can surpass classical computers. Exciting new areas at the intersection of quantum computing, communication, networked devices, and image analysis include Quantum Internet of Things (QIoT) and Quantum Image Processing. Using quantum technology, QIoT can enhance IoT safety, resource management, data analytics, and decision-making. Communication, sensing, and processing in the IoT are protected thanks to the use of quantum encryption, sensors, and machine learning algorithms.

The field of Quantum Image Processing examines and modifies pictures with the use of quantum computers. Quantum Computing's Quantum Fourier Transform-based image enhancement technique is highly effective for both low- and high-resolution images. The Quantum Fourier Transform has several other applications in satellite image processing. Methods for extracting quantum-inspired or quantum-based features from images improve such tasks as image segmentation, compression, enhancement, and pattern recognition. Improved accuracy and efficiency in image processing methods are possible with the help of quantum computing's parallelism and optimization.

Although QIoT and Quantum Image Processing have great potential, they are still in their infancy and face many obstacles. These include, but are not limited to, the creation of large-scale, fault-tolerant quantum computers and the incorporation of quantum technologies into existing IoT infrastructures. Improving QIoT and Quantum Image Processing calls for the creation of quantum hardware and algorithms.

QML shines when applied to smaller datasets for text and image classification issues. In the future, we hope to adapt some of the classical machine learning techniques for text categorization and investigate these issues from a purely quantum viewpoint. One of our future projects will be a discussion and identification of several outstanding issues and challenges in the field of quantum image processing, based on the state of the art in this area of study at the present time. This article provides useful guidance for the next generation of Internet of Things developers who seek to build solutions that are immune to the effects of quantum computing as a result of such a contribution.

REFERENCES

Coccia, M., Roshani, S., & Mosleh, M. (2022). Evolution of quantum computing: Theoretical and innovation management implications for emerging quantum industry. *IEEE Transactions on Engineering Management*, 1–11. doi:10.1109/TEM.2022.3175633

Dwivedi, A., Saini, G. K., & Musa, U. I. (2023, March). Cybersecurity and Prevention in the Quantum Era. In *2023 2nd International Conference for Innovation in Technology (INOCON)* (pp. 1-6). IEEE. 10.1109/INOCON57975.2023.10101186

Fernández-Caramés, T. M. (2019). From pre-quantum to post-quantum IoT security: A survey on quantum-resistant cryptosystems for the Internet of Things. *IEEE Internet of Things Journal*, *7*(7), 6457–6480. doi:10.1109/JIOT.2019.2958788

Islam, M., Chowdhury, M., Khan, Z., & Khan, S. M. (2022). Hybrid quantum-classical neural network for cloud-supported in-vehicle cyberattack detection. *IEEE Sensors Letters*, *6*(4), 1–4. doi:10.1109/LSENS.2022.3153931

Jadhav, A., Rasool, A., & Gyanchandani, M. (2023). Quantum Machine Learning: Scope for real-world problems. *Procedia Computer Science*, *218*, 2612–2625. doi:10.1016/j.procs.2023.01.235

Kalinin, M., & Krundyshev, V. (2023). Security intrusion detection using quantum machine learning techniques. *Journal of Computer Virology and Hacking Techniques*, *19*(1), 125–136. doi:10.100711416-022-00435-0

Kamaleswari, P., & Krishnaraj, N. (2023, February). An Assessment of Object Detection in Thermal (Infrared) Image Processing. In *2023 Third International Conference on Artificial Intelligence and Smart Energy (ICAIS)* (pp. 1498-1503). IEEE.

Khan, T. M., & Robles-Kelly, A. (2020). Machine learning: Quantum vs classical. *IEEE Access : Practical Innovations, Open Solutions*, *8*, 219275–219294. doi:10.1109/ACCESS.2020.3041719

Li, Y., Tian, M., Liu, G., Peng, C., & Jiao, L. (2020). Quantum optimization and quantum learning: A survey. *IEEE Access : Practical Innovations, Open Solutions*, *8*, 23568–23593. doi:10.1109/ACCESS.2020.2970105

Ralegankar, V. K., Bagul, J., Thakkar, B., Gupta, R., Tanwar, S., Sharma, G., & Davidson, I. E. (2021). Quantum cryptography-as-a-service for secure UAV communication: Applications, challenges, and case study. *IEEE Access : Practical Innovations, Open Solutions*, *10*, 1475–1492. doi:10.1109/ACCESS.2021.3138753

Ramachandra, A., & Singh, M. (2022). *A Hybrid Quantum-Classical Neural Network for Image Classification*. Innova Solutions.

Ranjan, A., Arya, A. K., & Ravinder, M. (2020, December). Quantum Techniques for Image Processing. In *2020 2nd International Conference on Advances in Computing, Communication Control and Networking (ICACCCN)* (pp. 1035-1039). IEEE. 10.1109/ICACCCN51052.2020.9362910

Rasool, R. U., Ahmad, H. F., Rafique, W., Qayyum, A., & Qadir, J. (2022). Quantum computing for healthcare. *RE:view*.

Ren, Z. A., Chen, Y. P., Liu, J. Y., Ding, H. J., & Wang, Q. (2020). Implementation of machine learning in quantum key distributions. *IEEE Communications Letters*, *25*(3), 940–944. doi:10.1109/LCOMM.2020.3040212

Sachdeva, K., Sachdeva, R., & Gupta, H. (2023). Quantum Computing in Image Processing. In Recent Developments in Electronics and Communication Systems (pp. 25-30). IOS Press. doi:10.3233/ATDE221232

Simões, R. D. M., Huber, P., Meier, N., Smailov, N., Füchslin, R. M., & Stockinger, K. (2023). Experimental Evaluation of Quantum Machine Learning Algorithms. *IEEE Access : Practical Innovations, Open Solutions*, *11*, 6197–6208. doi:10.1109/ACCESS.2023.3236409

Singh, A., Dev, K., Siljak, H., Joshi, H. D., & Magarini, M. (2021). Quantum internet—Applications, functionalities, enabling technologies, challenges, and research directions. *IEEE Communications Surveys and Tutorials*, *23*(4), 2218–2247. doi:10.1109/COMST.2021.3109944

Singh, J., & Bhangu, K. S. (2023). Contemporary Quantum Computing Use Cases: Taxonomy, Review and Challenges. *Archives of Computational Methods in Engineering*, *30*(1), 615–638. doi:10.100711831-022-09809-5

Srikanth, P., & Kumar, A. (2022). *Secure Quantum Computing for Healthcare Sector: A Short Analysis*. arXiv preprint arXiv:2211.10027.

Tacchino, F., Mangini, S., Barkoutsos, P. K., Macchiavello, C., Gerace, D., Tavernelli, I., & Bajoni, D. (2021). Variational learning for quantum artificial neural networks. *IEEE Transactions on Quantum Engineering*, *2*, 1–10. doi:10.1109/TQE.2021.3062494

Ullah, U., Jurado, A. G. O., Gonzalez, I. D., & Garcia-Zapirain, B. (2022). A Fully Connected Quantum Convolutional Neural Network for Classifying Ischemic Cardiopathy. *IEEE Access : Practical Innovations, Open Solutions*, *10*, 134592–134605. doi:10.1109/ACCESS.2022.3232307

Yang, Z., Zolanvari, M., & Jain, R. (2023). A Survey of Important Issues in Quantum Computing and Communications. *IEEE Communications Surveys and Tutorials*, *25*(2), 1059–1094. doi:10.1109/COMST.2023.3254481

Chapter 7
Quantum-Enabled Machine Learning With a Challenge in Clothing Classification With a QSVM Approach

Arvindhan Muthusamy
Galgotias University, India

A. Daniel
https://orcid.org/0000-0001-5564-2332
Amity University, India

ABSTRACT

Quantum information encoding also introduces the concept of quantum entanglement, where multiple qubits can become correlated in such a way that the state of one qubit cannot be described independently of the state of the other qubits. This property has important implications for quantum communication and quantum computing, as it allows for the creation of secure quantum communication channels and for the development of quantum algorithms that can solve certain problems much faster than classical algorithms. Quantum-enabled machine learning (QEML) is an emerging field that seeks to combine the power of quantum computing with classical machine learning techniques. One approach to QEML is the use of quantum support vector machines (QSVMs), which are quantum versions of classical support vector machines (SVMs).

DOI: 10.4018/978-1-6684-7535-5.ch007

Copyright © 2023, IGI Global. Copying or distributing in print or electronic forms without written permission of IGI Global is prohibited.

INTRODUCTION

For the purposes of carrying out computational operations, quantum computing is predicated on the utilisation of quantum mechanical phenomena such as superposition and entanglement. Quantum bits, that are employed in quantum computation, can exist in superpositions of states, in contrast to the binary digits (bits) that are used by classical computers, that can only be in one of two states (0 or 1) and cannot be in both states at the same time. In theory, these machines would be able to solve certain problems far more quickly than even the most modern classical computers could, if they used the most effective algorithms that are now available. An example of this would be the algorithm developed by Shor for factoring integers, and another would be the modelling of quantum many-body systems. Recently, for the very first time, this benefit, which is also referred to as "quantum supremacy," was demonstrated. There are primarily two different ways to perform quantum computing. To begin, there is a form of quantum computing known as gate-based computing, which is comparable to conventional digital computers. As a result of the difficulty involved in developing gate-based quantum computers, the majority of cutting-edge devices only have a limited quantity of qubits. The second technique is known as quantum annealing, and it was created by. It is projected that over the course of the next number of years, a quantum computer that is capable of being utilised in practise will be developed. In fewer than ten years from now, quantum computers will begin to overtake traditional computers, ushering in a new era that will see improvements in artificial intelligence, the creation of novel drugs, and other fields. Just a handful of the firms that are currently working on quantum processors include Google, IBM, Intel, Rigetti, QuTech, D-Wave, and IonQ. Other companies working on quantum chips include IBM, Intel, and others yet.

It is a well-established fact that quantum physics results in data patterns that are inconsistent with conventional wisdom. Deep neural networks, one of the most used approaches to machine learning, are a good example of a technique that consistently demonstrates the ability to not only recognise but also recreate statistical patterns in data. Following is an optimistic inference that can be drawn from this observation. A miniature quantum information processor might be able to recognise patterns that a traditional computer would have difficulty understanding due to their complex computational nature. For this dream to become a reality, it is absolutely necessary to achieve success in the development of efficient quantum algorithms for machine learning. Determining whether or not two graphs are isomorphic is an illustration of the type of challenge that a quantum algorithm is capable of resolving. Software for machine learning with a quantum twist incorporates quantum algorithms into a broader implementation as part of the software's overall design. Quantum algorithms have been proven to have the ability to outperform classical algorithms in some scenarios through careful evaluation of their prescribed operations. This has been shown to be the case. The term "quantum speedup" is used to refer to this potential benefit. Your starting point will determine whether you examine the concept of a quantum speedup from the viewpoint of formal computer science, which requires mathematical proofs, or from the viewpoint of what can be done with realistic, finite-size devices, which requires solid statistical evidence of a scaling advantage over some finite range of problem sizes. In either case, you will need to demonstrate that the quantum speedup provides a scaling advantage over some finite range of problem sizes. There are occasions when it is uncertain which classical algorithms will provide the optimum performance for quantum machine learning. Comparable to the circumstance surrounding Shor's polynomial-time quantum algorithm for integer factorization, in which a subexponential-time classical technique has not been discovered, but it has not been established beyond a reasonable doubt that such a technique cannot exist. For the benchmarking challenge to be solved, which compares the scaling of quantum machine learning to that of classical machine learning, we require quantum computers. There is a possibility of benefits, such as greater quality classifications and data sampling from systems that were unavailable earlier. As a direct consequence of this, idealised metrics derived from complexity theory, such as query complexity and gate complexity, are currently being utilised in order to characterise quantum speedups in machine learning. The number of requests that a conventional or quantum algorithm needs to make to the primary data source is referred to as the query complexity of the algorithm. If a quantum approach can answer a problem with a smaller number of queries than a classical one can, then it is more efficient than the classical method. In order to determine the gate complexity, count the number of "gates," also known as elementary quantum operations. These are required in order to achieve the result that is desired. Quantifying the resources needed to solve a problem class can be done with the help of idealized models such as query and gate complexity. There isn't much that can be said about the required scaling of resources in a situation that takes place in the real world before first figuring out how to translate this idealization into the real world. As a direct consequence of this, numerical testing is the major method that is used for assessing the resources that are required by classical machine learning algorithms. In practice, it is likely to be difficult to quantify the amount of resources that quantum machine learning algorithms use. In this analysis, our primary focus is on determining whether or not their implementation in real-world settings is practical.

Figure 1. Quantum methods for machine learning

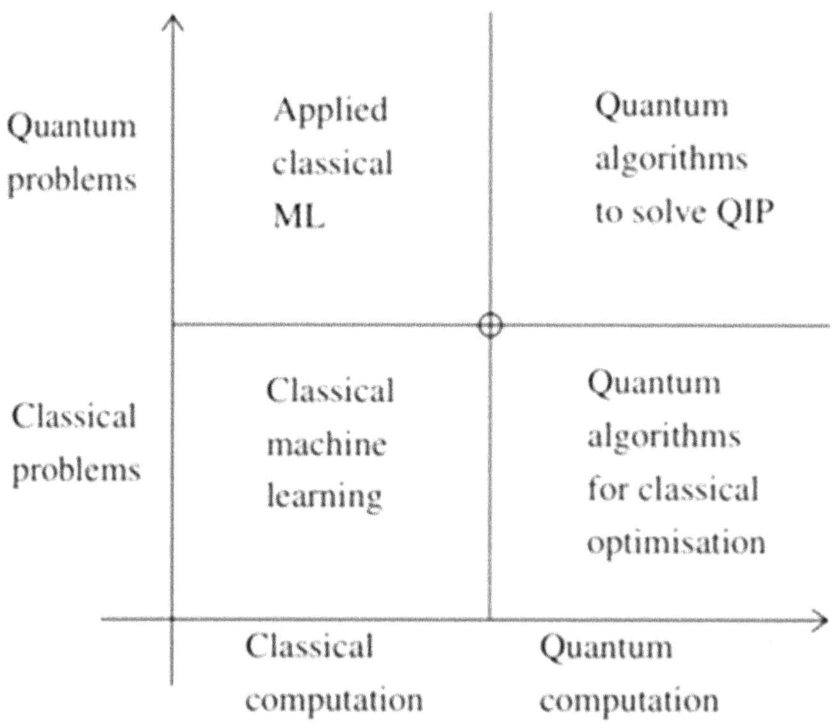

There are quantum methods for machine learning that have been shown to demonstrate quantum speedups as will be observed during the course of this review. For instance, Fourier transformations, locating eigenvectors and eigenvalues, and solving linear equations all display exponential quantum speedups in comparison to their best known classical equivalents (Giovannetti et al., 2008; Kimmel et al., 2016; Le, 2013). This quantum BLAS (qBLAS) translates into quantum speedups for a variety of algorithms used for data analysis and machine learning. Some of these algorithms include linear algebra, least-squares fitting, gradient descent, Newton's method, principal component analysis, linear, semidefinite, and quadratic programming, topological analysis, and support vector machines. At the same time, deep learning architectures are a good fit for special purpose quantum information processors. Some examples of these processors include quantum annealers and programmable quantum optical arrays. In spite of the fact that it is not quite known as of yet to what extent this potential is born out in reality, there are grounds to be optimistic that quantum computers will be able to recognise patterns in data that classical computers will not be able to. The following is how we have organised this review. Classical or quantum computers can be used to learn. There is a possibility that the data they analyse are of a classical or quantum state, and that it was produced by quantum sensing or measuring device. Conventional machine learning, also known as the use of traditional computers to identify patterns in traditional data, is briefly covered here.

CLASSICAL MACHINE LEARNING

Traditional approaches to machine learning and statistical analysis can be broken down into a few distinct areas. To begin, computers can be used to carry out 'traditional' methods of data analysis, such as least squares regression, polynomial interpolation, and data analysis. Both supervised and unsupervised approaches to machine learning are possible. During the supervised learning process, the training data is separated into labelled categories, such as examples of handwritten digits together with the actual number that each handwritten digit is supposed to represent. The machine's task is to learn how to assign labels to data that is not part of the training set. Unsupervised learning is a sort of machine learning in which the training set is not labelled. The purpose of this type of learning is for the machine to discover the natural categories into which the training data falls (for example, the various types of images that can be found on the internet), and then to classify data that is not part of the training set. Last but not least, there are applications of machine learning, such as playing the game Go, that require a mix of supervised and unsupervised learning, in addition to training sets that may be generated by the machine itself.

QUANTUM MACHINE LEARNING

Quantum computing could have some potentially fascinating applications, one of which is machine learning. The current classical methods call for enormous amounts of computer resources, and in many instances, training takes a significant amount of time. The user or programmer does not give the machine any explicit instructions; rather, the machine learns on its own through experience and the usage of data samples. This type of learning is referred to as machine learning. When considering traditional machine learning, one can differentiate between the following types:

Supervised learning: in which labelled data are utilised, for example while dealing with classification issues. This indicates that the data that is utilised for the purpose of learning has information regarding the category to which it belongs. Learning without supervision involves making use of unlabeled data in order to solve problems such as grouping. In this case, data points need to be assigned to a certain cluster of like points, but there is no prior information available.

Semi-supervised learning: in this type of learning, only some of the data are labelled, and researchers examine different models to see how they might improve classification by combining labelled data with unlabeled data. A good number of these models make use of probabilistic and generative methodologies.

Reinforcement learning: is a type of machine learning in which there is no labelled data available; however, a technique is employed to measure the computer's success in the form of incentives. The machine explores a wide variety of potential courses of action and determines the most effective ones by analysing the responses (rewards) it receives after completing each one. The possibility for quantum computers to perform computations more quickly is one of the most significant advantages of using these machines. When compared to classical algorithms, quantum algorithms can achieve a speed-up that is either polynomial or super-polynomial (exponential), depending on the nature of the problem and the algorithm. On the other hand, we anticipate that other benefits will become more significant in the not too distant future. It's possible that quantum computers may learn from less input, handle more complex structures, or be better at managing with noisy data. In a nutshell, the three primary advantages of utilising quantum machine learning are as follows (interpretation based on (Lloyd, 1996)): Improvements in run-time: obtaining faster results; Learning capacity improvements: increase of the capacity of associative or content addressable memories; Understanding efficiency enhancements: less training knowledge or simpler algorithms needed to produce the same outcomes or more complicated relations can be understood from the identical data.

Figure 2. A quantum neural network consists of a layered structure with parameter-dependent unitary operations. The lines correspond to qubits, with the lowest one being the readout qubit and the other ones being the data qubits.

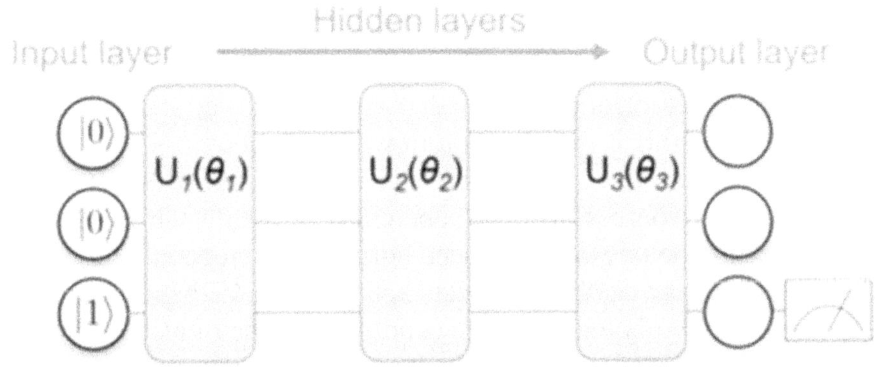

LINEAR-ALGEBRA BASED QUANTUM MACHINE LEARNING

Matrix operations on vectors in a high-dimensional vector space are used in a broad variety of data analysis and machine learning protocols. These protocols are used to execute their respective functions. Quantum mechanics, on the other hand, is all about performing matrix operations on vectors while in high-dimensional vector spaces. When quantum logic operations or measurements are conducted on qubits, the associated state vector is multiplied by 2n 2 n matrices. This is the essential component that underpins these methods. Quantum bits, also known as qubits, each have a quantum state that is represented as a vector in a complex vector space that has 2n dimensions. Quantum computers have been shown to be capable of performing common linear algebraic operations such as Fourier transforms, finding eigenvectors and eigenvalues, and solving linear sets of equations over 2n-dimensional vector spaces in time polynomial in n, which is exponentially faster than their best known classical counterparts. This is accomplished by building up such matrix transformations on a quantum computer. This latter approach is what the authors of the paper refer to as the HHL algorithm (see Box 2 for further information). The first version of this type made the assumption of a well-conditioned yet sparse matrix. In the field of data science, sparsity is extremely improbable; yet, complying with developments broadened the scope of this assumption to encompass low-rank matrices. Moving beyond HHL, the purpose of this article is to present an overview of many quantum algorithms that manifest themselves as subroutines whenever linear algebra methods are used in quantum machine learning software.

Unpredictable quantum objects such as photons that originate from a source of unpolarized photons or that result from the combination of photons with different polarity, as seen in the image on the right of the following paragraph. At this moment, the polarisation of the photons can be thought of as a statistical combination of photons with horizontal and vertical polarisation. Let's imagine that this is the situation in which quantum objects that have been created in diverse ways are subsequently combined with one another. The two different sources do not "coherently" present their information. When applied to photons that have been horizontally prepared, the example on the left shows how a polarising beam splitter set at 45 degrees can create superposed H and V photons in a pure condition. On the right, the polarising beam splitter generates photons that are horizontally and vertically polarised at a ratio of 50/50. These photons can then be merged by a non-polarizing beam splitter that is rotated 45 degrees. They are combined statistically, but they are not layered atop one another, which results in a mixed state.

Figure 3. Quantum principal component

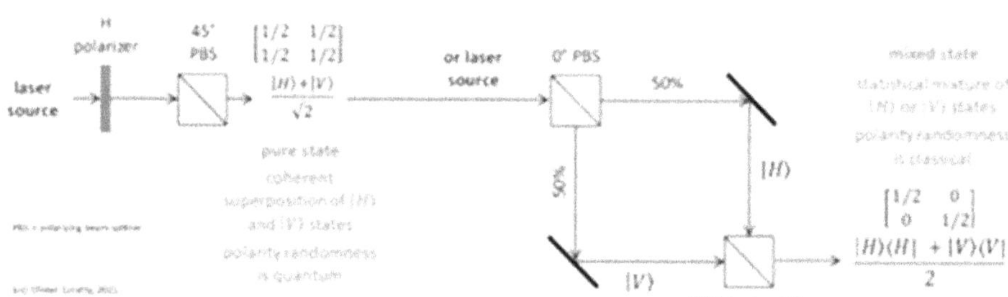

QUANTUM PRINCIPAL COMPONENT ANALYSIS

Quantum principal component analysis of classical data (qPCA) requires us to pick a data vector vj at random and then use a quantum random access memory (qRAM) to transfer that vector into a quantum state. qPCA stands for quantum principal component analysis of classical data. ~vj ® |vj i. The quantum state that summarises the vector contains log d qubits, and the operation of the qRAM requires O(d) operations that are divided across O(log d) steps that can be carried out in parallel with one another. The quantum state that was produced as a result of the random selection of vj has a density matrix that looks like this: = (1/N) P j |vj ihvj |, where N is the number of data vectors. When we compare the covariance matrix C for the classical data with the density matrix for the quantum version of the data, we notice that the density matrix is the same as the covariance matrix, up to an overall factor. The properties of C's primary components can then be investigated by making measurements on the quantum representation of C's eigenvectors. This will allow for a more complete understanding of C's structure. The quantum algorithm has a computational complexity that scales as O (log d) 2 and a query complexity that scales as O (log d) 2. To put it another way, the quantum PCA is orders of magnitude more effective than the classical PC.

Figure 4. Machine learning

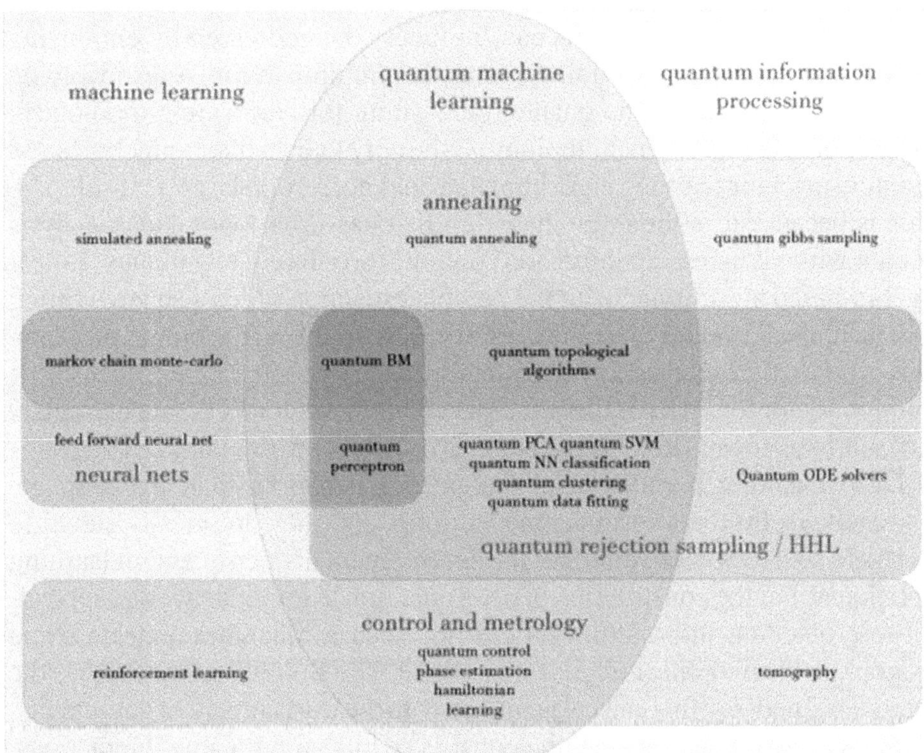

READING CLASSICAL DATA INTO QUANTUM MACHINES

In order for a quantum computer to process data, it first requires the input of classical information. This is something that is frequently referred to as "the input problem," and while it is frequently accomplished with very little overhead, it does cause a significant bottleneck for certain algorithms. When reading out data that has been processed on a quantum device, the 'output problem' also arises as a challenge. The output difficulty, much like the input problem, frequently brings about a major delay in operational speed. To begin the process of applying HHL, the method of least squares fitting, qPCA, quantum support vector machines, and other similar methods to classical data, the operation begins with initially loading substantial amounts of data into a quantum system, which can need exponential time . This is especially true if we consider applying HHL, least squares fitting, qPCR, and other related methods to classical data. This can, in theory, be addressed by employing qRAM, but the expense of doing so may be prohibitive for issues involving large amounts of data. The quantum algorithm for performing topological analysis of data (persistent homology) is the only known linear-algebra based quantum machine learning algorithm that does not rely on large-scale qRAM. This is because it is the only algorithm that can do a topological analysis of data without using combinatorial optimization-based techniques. Linear algebra-based algorithms, with the notable exception of least squares fitting and quantum aid vector machines, are also susceptible to the output problem. This is due to the fact that classical quantities that are sought after, such as the solution vector for HHL or the principal components for PCA, are extremely difficult to estimate (Rebentrost et al., 2014; Rebentrost et al., 2016).

Deep quantum learning Classical neural networks that are deep are extremely useful tools for machine learning, and they are ideally suited to serve as a source of motivation for the development of deep quantum learning techniques. For the construction of deep quantum learning networks, special-purpose quantum information processors, such as quantum annealers and programmable photonic circuits, are optimal. The Boltzmann machine is the deep neural network that can be quantized with the least amount of complexity. The classical Boltzmann machine consists of bits with tunable interactions. The Boltzmann machine is trained by modifying those interactions in such a way that the thermal statistics of the bits, which are defined by a Boltzmann Gibbs distribution, reproduce the statistics of the data. This allows the Boltzmann machine to accurately represent the data. To quantize the Boltzmann machine, one need just take the neural network and describe it as a set of interacting quantum spins, which corresponds to a tunable Ising model. This is all that is required to complete the quantization process. Then, in order to acquire a response, we can read out the output qubits by setting the starting state of the input neurons in the Boltzmann machines to a predetermined value and allowing the system to reach its thermal equilibrium. One of the most important aspects of deep quantum learning is the fact that it does not necessitate the use of a huge quantum computer designed for all purposes. Quantum annealers are a specialised type of quantum information processor that, in comparison to general-purpose quantum computers, are substantially less complicated to build and more straightforward to expand . Quantum annealers are readily available on the market and work exceptionally well for the implementation of deep quantum learning systems. The D-Wave quantum annealer is a programmable model of transverse Ising that may be tuned to produce thermal states of classical as well as specific quantum spin systems. Deep quantum learning algorithms have been successfully carried out on more than a thousand spins by means of the D-Wave device (Rosenblatt, 1958; Scherer, 2015; Schuld et al., 2016). Quantum Boltzmann machines with more general tunable couplings are currently in the design stage. These machines will be able to implement universal quantum logic once they are completed. On-chip silicon waveguides have been utilized in the construction of linear optical arrays comprising hundreds of tunable interferometers. Additionally, special purpose superconducting quantum information processors have the potential to be utilized in the implementation of the QAO algorithm. Quantum computers have a number of potential applications that could be advantageous in this context. To begin, using quantum approaches can cause a system to reach its thermal equilibrium four times as quickly as it would using its classical counterpart . Because of this, it may become possible to conduct correct training on fully networked Boltzmann machines. Second, because they offer more effective methods of sampling, quantum computers can speed up the Boltzmann training process. Because of the random nature of the neuron activation pattern in the Boltzmann machine, a significant number of iterations are required in order to determine success probability and, consequently, the impact that modifying a weight in the neural network has on the overall performance of the deep network. Quantum coherence, on the other hand, can cut the number of training samples required for a quantum Boltzmann machine by a factor of four. This helps the system learn its performance more quickly. A quantum algorithm can train a deep neural network on a large training data set while only reading a minuscule number of training vectors. This is possible because quantum access to the training data (i.e., qRAM or a quantum blackbox subroutine) enables the machine to be trained using quadratically fewer accesses to the training data than is required by classical methods. Deep learning can benefit from the introduction of new essentially quantum models thanks to quantum information processing. For instance, by including a transverse field in the basic Ising model quantum Boltzmann machine described above, one can obtain a transverse Ising model, which is capable of displaying a wide range of fundamental quantum effects, such as entanglement . The quantum Boltzmann machine can be transformed into a wide range of different quantum systems . All it takes is the addition of a few more quantum connections. Adding an adjustable transverse interaction to a tunable Ising model is known to be universal for full quantum computing (Farhi et al., 2014; Whitfield et al., 2012). Provided that the appropriate weight assignments are made, this model is capable of executing every algorithm that can be performed by a general-purpose quantum computer.

QML DATASETS

In spite of the fact that quantum information for machine learning (both quantum and classical) are not nearly as common or as pervasive as their classical counterparts, there are some examples in the scientific literature of quantum or hybrid quantum-classical datasets that were created for applications in machine learning contexts. These datasets can be used to train machine learning models. Quantum datasets that were created for purposes other than QML, such as quantum datasets in quantum chemistry or other domains, and that can be preprocessed or utilized as training information in QML settings can be categorized as follows: (1) general quantum datasets developed for purposes other than QML. These datasets were not generated for the purposes of QML per se when they were created; (2) dedicated QML-specific quantum datasets, which were prepared and structured for the purposes of QML. The majority of the datasets in this second category are quantum-based, but they are designed for usage in traditional or hybrid machine-learning settings. Quantum datasets that are now accessible often fall into one of these two classes, however there is overlap between them. For instance, quantum datasets developed for use in machine learning are extremely domain-specific despite the fact that there is some overlap between them. For instance, there are datasets for quantum chemistry that may be utilised in deep tensor neural networks datasets for learning the spectrum features of molecular systems and datasets for solid-state physics.

The QM7-X dataset covers various sections of the space occupied by chemical compounds. It was produced to serve as a foundation for the machine-learning-assisted design of molecules with particular characteristics. This dataset is an expansion of earlier iterations of the QM-series and other quantum chemistry-related datasets. The ground state qualities (spectra and moments) along with the response values (related to polarisation and dispersion) are included in the dataset. Structurally, the dataset combines global (molecular) attributes and local (atomic) information. The dataset has a high level of domain specificity and is a prominent example of a dataset that was developed to encourage machine learning driven research within a given discipline.

QUANTUM MACHINE LEARNING FOR QUANTUM DATA

Quantum data, or the states produced by quantum systems and processes, may provide the most direct application for quantum machine learning. Many quantum machine learning algorithms, like those mentioned above, use a mapping from classical data to quantum mechanical states and then manipulate those states with elementary quantum linear algebra subroutines in order to discover patterns in the original data. When applied to the quantum states of light and matter, these same machine learning methods uncover previously hidden characteristics and patterns. In many cases, the quantum modes of analysis that emerge from studying quantum systems are far superior to their classical counterparts. In contrast to the O(N2) measurements and O(N) operations required for a classical device to perform tomography on a density matrix and the O(N) operations required for the classical PCA, quantum principal component analysis can find its eigenvalues and reveal the corresponding eigenvectors in time O((log2 N) 2). Smaller quantum computers are expected to become commercially available in the next few years, and they might be used to execute such quantum analysis of quantum data. Quantum simulators, when used to investigate quantum dynamics, are a potent tool for quantum data processing. Quantum simulators are essentially “quantum analogue computers,” or quantum systems whose dynamics can be tailored to mimic that of a specific system of interest. The term “quantum simulator” can refer to either a general-purpose quantum computer or a dedicated device built to model a specific type of quantum system. Connecting a reliable quantum simulator to an unknown system and adjusting the simulator’s model to account for the unknown dynamics allows for fast learning of the unknown system's dynamics using approximate Bayesian inference. As a result, the simulation requires significantly fewer measurements. The quantum Boltzmann training algorithm of allows states to be reconstructed in time logarithmic in the dimension of the Hilbert space, which is an exponential speedup over classical tomography for reconstructing quantum dynamics. The loading of coherent input states is a significant technological barrier for using a quantum computer to aid in the characterization of a quantum system or for use in a quantum PCA algorithm. However, such applications remain among the interesting prospects for near-term use of quantum machine learning because they do not require QRAM and offer the possibility for exponential speedups for device characterization.

DESIGNING AND CONTROLLING QUANTUM SYSTEMS

Tuning quantum gates to meet the stringent requirements of quantum error correction is a significant obstacle to the advancement of quantum computation and information technology. For example, in the case of nearest-neighbor coupled superconducting artificial atoms with gate fidelity above 99.9% in the presence of noise, heuristic search methods can help achieve this in a supervised learning scenario, meeting a widely accepted threshold for fault-tolerant quantum computing. Using the same approach, a single-shot Toffoli gate was built with a fidelity of over 99.9 percent. The ability to exert control over a quantum system is equally as crucial and intricate. Success with learning methods has also been achieved in optimizing adaptive quantum metrology, a fundamental component of many quantum technologies. Some researchers have proposed using genetic algorithms for quantum molecule control as a way to account for fluctuations in experimental conditions. Especially in the presence of noise and decoherence, reinforcement learning algorithms that use heuristic global optimization, like the one in circuit design, have proven to be highly effective. Quantum systems built on gates can also benefit from reinforcement learning. When exposed to an external stray field of undetermined magnitude and direction, adaptive controllers based on intelligent agents for quantum information display adaptive calibration and compensation mechanisms. To get theoretical insights into quantum states, classical machine learning is also a potent tool. Recently, neural networks have been used to investigate two fundamental issues in condensed matter: the detection of phase of matter and the search for the ground state. They have a lot of success getting superior results to standard numerical software.

Theoretical physicists are currently analyzing these models to learn how well they describe reality in comparison to more conventional approaches like tensor networks. It has been demonstrated that these applications may capture highly nontrivial aspects from disordered or topologically ordered systems, making them a promising tool for studying unusual states of matter.

PERSPECTIVES ON FUTURE WORK

This paper demonstrates the promising uses of both small quantum computers and bigger special purpose quantum simulators, annealers, etc., in the fields of machine learning and data analysis. These algorithms can only be run on quantum technology, hence the question is whether or not their potential will ever be realised. Several enabling technologies have made significant advancements on the hardware side. Quantum cloud computing (the 'Qloud') will make small-scale quantum computers with 50-100 qubits broadly available. Quantum information processors designed for a specific task will continue to grow in power and sophistication. These include quantum simulators, quantum annealers, integrated photonic chips, NV-diamond arrays, quantum random access memory, and custom-built superconducting circuits. Applications in quantum machine learning for compact quantum computers are bolstered by the inclusion of specialised quantum information processors digital quantum processors and quantum sensors.In particular, scalable integrated superconducting circuits have been used to construct and run quantum annealers with 2000 qubits. In order to implement quantum machine learning algorithms, quantum annealers must overcome significant hurdles, such as enhancing connectivity and implementing more generally tunable couplings between qubits. Using integrated photonics in silicon, programmable quantum optic arrays with a few hundred tuneable interferometers have been built, although loss remains a significant barrier to further developing and commercializing these kinds of circuits. The development of interface devices such quantum random access memory (qRAM) that permit the encoding of classical information in quantum mechanical form (Arunachalam et al., 2015) represents a significant challenge for quantum machine learning. A quantum random access memory (qRAM) that can retrieve N bits of information has 2N quantum switches that must all be in perfect sync in order to retrieve the requested data. Memory access in such a qRAM requires O(log2 N) time and it can accept error rates of up to O(1/ log2 N) per switching operation, where N is the depth of the qRAM. While qRAM proof-of-concept demonstrations exist, building massive arrays of quantum switches remains a challenging technological challenge. These hardware issues are purely technical in nature, with well-defined solutions. In order for quantum machine learning to be the 'killer app' for quantum computers, certain challenges must be surmounted. As was said before, most of the known quantum algorithms have a variety of restrictions that prevent them from being widely used. We can summarise the aforementioned qualifiers into four primary issues. First, there's the input issue; while quantum algorithms can greatly accelerate processing times, they're not usually better at reading data. The time spent reading the input may end up being more expensive than running the actual quantum computation. Research is ongoing to further understand this potential protective mechanism. The output is that some quantum algorithms have an exponentially large number of bits to learn in order to learn the whole solution as a bit string. As a result, several uses of QML algorithms become impractical. Learning only the solution state's summary statistics could be a workaround for this issue, however this is still a work in progress. Thirdly, the issue of cost is closely tied to input/output issues, as it is unknown at this time how many gates are necessary for quantum machine learning algorithms. Complexity bounds imply they will provide significant benefits for problems of a certain size, but when this occurs is still an open subject. Issues with Benchmarking 4. Due to the substantial benchmarking against modern heuristic approaches that would be required, it is generally difficult to say that a quantum algorithm is ever superior than all known classical machine algorithms in practise. This could be somewhat fixed by more findings setting lower constraints for quantum machine learning. One possible way forward that avoids some of these problems is to investigate how quantum computing can be used to quantum data as opposed to classical data. Quantum machine learning is being used to understand and manage quantum computers. This would pave the way for a virtuous cycle of invention, analogous to what took place in classical computing, in which the strengths of one generation of processors were used to develop the next. The first fruits of this cycle have already begun to appear, with classical machine learning being used to better the design of quantum processors, which in turn provide potent computational resources for quantum-enhanced machine learning applications.

CONCLUSION

The field of machine learning has seen substantial growth and development in recent years due to the emergence of quantum computing. Quantum-enabled machine learning (QEML) is a novel approach that combines the capabilities of classical machine learning algorithms with the power of quantum computation. One particular challenge in this area is clothing classification, which involves identifying different types and styles of clothing items. To address this issue, researchers have proposed a QEML-based solution using a quantum support vector machine (QSVM) approach. The study involved training and testing on an image dataset consisting of clothing images belonging to ten categories. Results showed that the QSVM outperformed classical SVMs in terms of accuracy for classifying clothing types. In addition, privacy preservation was also an important factor considered by the researchers during their analysis. They addressed this concern through anonymization techniques rather than selectively releasing information based on privacy preferences enforcement. This research highlights the potential usefulness and superiority of QEML approaches over traditional methods for tackling complex problems such as clothing classification. Furthermore, it demonstrates how integrating blockchain technology into IoT architecture could enhance confidentiality and security measures while maintaining data sharing abilities among shared economic applications and beyond.

REFERENCES

Arunachalam, S., Gheorghiu, V., Jochym-O'Connor, T., Mosca, M., & Srinivasan, P. V. (2015). On the robustness of bucket brigade quantum RAM. *New Journal of Physics*, *17*(12), 123010. doi:10.1088/1367-2630/17/12/123010

Farhi, E., Goldstone, J., & Gutmann, S. (2014). *A quantum approximate optimization algorithm*. https://arxiv.org/abs/1411.4028

Giovannetti, V., Lloyd, S., & Maccone, L. (2008). Quantum random access memory. *Physical Review Letters*, *100*(16), 160501. doi:10.1103/PhysRevLett.100.160501 PMID:18518173

Kimmel, S., Lin, C. Y.-Y., Low, G. H., Ozols, M., & Yoder, T. J. (2016). *Hamiltonian simulation with optimal sample complexity*. https://arxiv.org/abs/1608.00281

Le, Q. V. (2013). Building high-level features using large scale unsupervised learning. In *Acoustics, Speech and Signal Processing (ICASSP), IEEE International Conference on,* (pp. 8595–8598). IEEE. 10.1109/ICASSP.2013.6639343

Lloyd, S. (1996). Universal quantum simulators. Science 273, 1073–1078. doi:10.1126cience.273.5278.1073

Rebentrost, P., Mohseni, M., & Lloyd, S. (2014). Quantum support vector machine for big data classification. *Physical Review Letters*, *113*(13), 130503. doi:10.1103/PhysRevLett.113.130503 PMID:25302877

RebentrostP.SteffensA.LloydS. (2016). *Quantum singular value decomposition of non-sparse low-rank matrices*. https://arxiv.org/abs/1607.05404

Rosenblatt, F. (1958). The perceptron: A probabilistic model for information storage and organization in the brain. *Psychological Review*, *65*(6), 386–408. doi:10.1037/h0042519 PMID:13602029

Scherer, A. (2015). *Resource analysis of the quantum linear system algorithm*. https://arxiv.org/abs/1505.06552

Schuld, M., Sinayskiy, I., & Petruccione, F. (2016). Prediction by linear regression on a quantum computer. *Physical Review. A*, *94*(2), 022342. doi:10.1103/PhysRevA.94.022342

Whitfield, J. D., Faccin, M., & Biamonte, J. D. (2012). Ground-state spin logic. *Europhysics Letters*, *99*, 57004. doi:10.1209/0295-5075/99/57004

Chapter 8
Quantum Machine Learning Enhancing AI With Quantum Computing

Arvindhan Muthusamy
Galgotias University, India

ABSTRACT

Quantum circuits with certain statistical properties called t-designs allow investigations into the behavior of gradients in high dimensions, and therefore tell us something about the trainability of quantum models in regimes that cannot necessarily be simulated. Along with that, the chapter shows how quantum machine learning is being used in real-world settings in a variety of fields, such as drug discovery, financial modelling, image identification, and natural language processing. All fundamental subatomic particles fall into two classes, based on how identical particles of each type behave when swapped. They are either fermions, a class that includes electrons and other particles that make up matter, or bosons, which include particles of light known as photons. Data encoding is often the most crucial step of quantum machine learning with classical data. It illustrates the possible advantages and insights that quantum algorithms can offer to various fields and analyses the ongoing research projects aimed at converting conventional machine learning models for quantum platforms.

DOI: 10.4018/978-1-6684-7535-5.ch008

Copyright © 2023, IGI Global. Copying or distributing in print or electronic forms without written permission of IGI Global is prohibited.

1. INTRODUCTION

In the fast developing subject of quantum computing, computation is done using the concepts of quantum mechanics. It makes use of quantum bits, also known as qubits, which can be in several states at once due to the concept of superposition. As a result, quantum computers can handle enormous volumes of information at once and might even be able to solve some problems more quickly than traditional computers. Entanglement, in which qubits are correlated, and quantum gates, which change qubit states, are important ideas in quantum computing. The goal of artificial intelligence (AI) is to create intelligent machines with capabilities equivalent to those of human beings in terms of perception, learning, reasoning, and decision-making. In order to teach algorithms, which are a subset of AI, how to recognize patterns and make predictions or judgements, massive datasets are used in machine learning. Deep learning, a kind of machine learning, processes complex data and extracts useful features using neural networks with numerous layers. Natural language processing, computer vision, robotics, recommendation systems, and other technologies are just a few of the many uses for AI.

The potential for quantum computing and AI to improve and complement one another is what binds them together. The enormous computational power and simultaneous data processing capacity of quantum computing has the potential to speed up and improve AI algorithms. Quantum algorithms created expressly for AI activities to address optimization issues, better pattern recognition, and improve data processing can use the special characteristics of quantum systems. Moreover, conventional AI algorithms may be improved and quantum-inspired algorithms, which are inspired by quantum principles but can be implemented on classical computers, may overcome some of its drawbacks. A tool for AI research, quantum computing enables researchers to explore novel architectures, simulate and analyse sophisticated AI models, and create cutting-edge methods that push the field's limits. The relationship between quantum computing and AI is still not fully understood, but it shows promise for addressing difficult issues, enhancing computational effectiveness, and expanding the capabilities of intelligent systems. Unlocking the transformational potential of this connection depends on ongoing research and cooperation between the communities of quantum computing and AI.

The domains of machine learning and quantum computing have each made outstanding advancements recently, changing the technological landscape and creating new opportunities for innovation. Using the laws of quantum physics, quantum computing has the potential to tackle complicated problems tenfold more quickly than traditional computers. A revolution in artificial intelligence has been brought about by machine learning, which enables computers to learn from data and make wise predictions or judgements. Now that these two domains have converged, a new field called quantum machine learning (QML) has emerged. QML aims to use quantum computers to improve AI skills and realise the quantum advantage. QML has a lot of potential to revolutionize a number of AI-related fields, including pattern identification, optimization, and data analysis. QML algorithms have the potential to surpass their classical counterparts in terms of computational correctness and efficiency by taking advantage of the special characteristics of quantum systems, such as superposition and entanglement. Researchers and industry professionals alike have paid close attention to the interaction between quantum computing and machine learning, which has led to fascinating developments and discoveries.

In this article, we investigate the field of quantum machine learning and examine how the computational power of quantum computing might be used to improve AI. We will go over the fundamental ideas, procedures, and methods that underlie QML in order to highlight its revolutionary potential for a range of applications. We will also look at the difficulties and factors that come with combining machine learning with quantum computing, such as algorithm design, hardware limitations, and the creation of specialized quantum software frameworks. We hope to give a thorough overview of this developing topic and highlight its significance for the development of artificial intelligence through this investigation of quantum machine learning. We can unleash the full potential of machine learning and expand the frontiers of artificial intelligence by utilizing the power of quantum computing.

2. CONTRASTING CLASSICAL MACHINE LEARNING WITH QUANTUM MACHINE LEARNING: UNDERSTANDING THE FUNDAMENTAL DIFFERENCES

The primary distinction between classical machine learning (ML) and quantum machine learning (QML) is found in the computational models that underlie each approach, as well as the methods for processing and analyzing data. Classical machine learning employs a deterministic and sequential execution model and runs on traditional computers. Information is represented and processed using traditional bits. Traditional machine learning (ML) methods are made to operate with data that is often represented as numerical values or categorical labels. These algorithms make use of statistical techniques, mathematical algorithms like gradient descent, decision trees, and support vector machines, as well as traditional optimization techniques. Since classical ML has been thoroughly researched and optimized for classical computer architectures, it can be used to solve a variety of real-world issues.

Figure 1. Contrasting classical machine learning with quantum machine learning

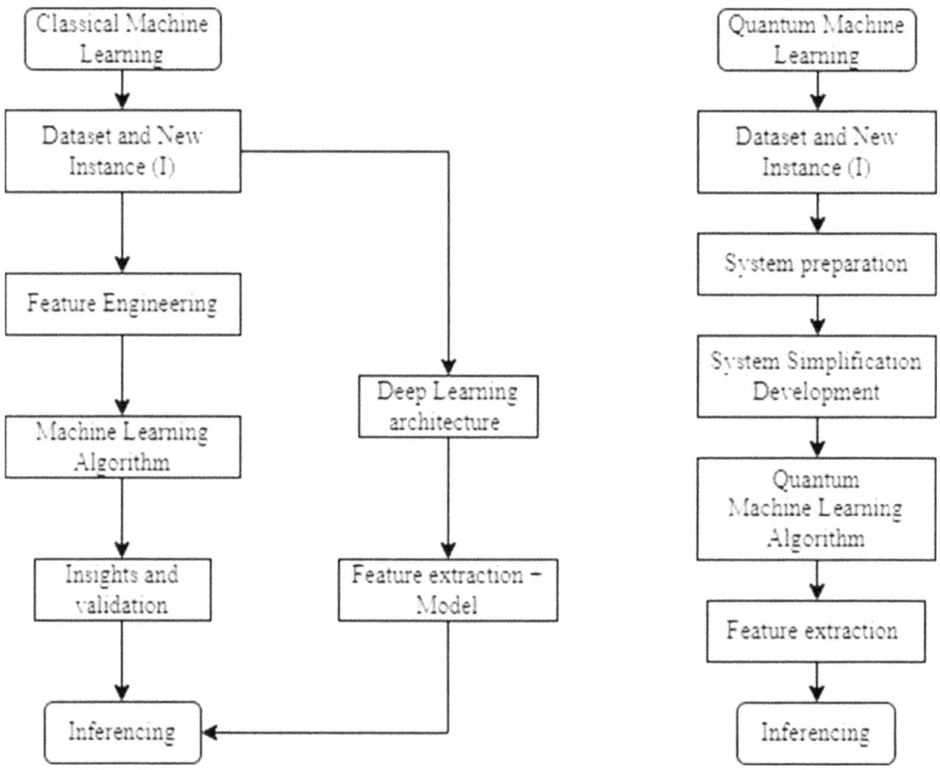

On the other hand, QML makes use of quantum computing principles to improve machine-learning capabilities. Qubits, which can represent and process information in quantum states, are the building blocks of quantum computers. Qubits, which differ from traditional bits in that they can reside in a superposition of 0 and 1, allow for parallel processing. It is the purpose of QML algorithms to take use of the special qualities of quantum systems. To compute on quantum states, they use quantum gates and quantum circuits. Quantum versions of conventional machine learning algorithms as well as original algorithms created especially for quantum systems are both included in QML algorithms. These algorithms may be able to exponentially speed up specific tasks like data classification, optimization, and pattern recognition by taking advantage of quantum phenomena like superposition and entanglement. Both classical and quantum data can be included in the data representation in QML. Quantum states can be used to encode classical data, enabling parallel processing and possible speedup. Contrarily, quantum data is information that can be processed directly by quantum algorithms since it is fundamentally quantum in nature. It's crucial to remember that real-world QML applications are still developing since it's difficult to create viable quantum computers with enough qubits and low error rates.

3. QUANTUM POWER HARNESSED FOR ADVANCED MACHINE LEARNING

In many domains, including machine learning, new frontiers have been made possible by the quick development of quantum computing. The use of quantum computing allows for the performance of intricate calculations and the resolution of issues that are beyond the capabilities of conventional computers. Researchers have looked into the possibility of using quantum computing for sophisticated machine learning tasks in recent years. In addition to highlighting the key ideas and methods used in utilizing quantum computing for improved machine learning capabilities, this paper provides an overview of the developing topic of quantum machine learning. We explore quantum algorithms that have the potential to enhance pattern recognition, optimization, and data processing tasks, such as quantum support vector machines and quantum neural networks. Also covered are the difficulties and possibilities of quantum machine learning, such as hardware limitations, algorithmic design, and the requirement for specialized quantum software development. We hope to stimulate additional research and development in this fascinating field and open the door for revolutionary developments in artificial intelligence by investigating the synergy between quantum computing and machine learning.

4. COMBINING MACHINE LEARNING AND QUANTUM COMPUTING TO UNLEASH THE QUANTUM ADVANTAGE IN AI

By utilizing the quantum advantage, machine learning and quantum computing have the potential to revolutionize the field of AI. The capabilities of classical computers are surpassed by quantum computing, which applies the ideas of quantum mechanics to computation. Machine learning algorithms provide the ability for computers to learn from data and make predictions or judgements. Combining these two potent technologies allows us to take use of the special qualities of quantum systems and improve the efficacy and efficiency of AI applications. In this article, we investigate the relationship between machine learning and quantum computing, emphasizing crucial applications of quantum advantage in AI. We examine quantum-inspired algorithms, quantum neural networks, and quantum-enhanced optimization methods, demonstrating their capability to speed up training procedures, boost pattern recognition, and resolve challenging optimization issues. We also explore the difficulties and factors involved in combining machine learning and quantum computing, including algorithm design, hardware limitations, and the requirement for specialized quantum software frameworks. We can advance the field of artificial intelligence by embracing the merger of machine learning with quantum computing, which will result in innovations in a number of industries, including banking, healthcare, and natural language processing.

5. INVESTIGATING THE RELATIONSHIP BETWEEN QUANTUM COMPUTING AND AI

Both quantum computing and artificial intelligence (AI) have the potential to have significant effects on one another. While quantum computing and AI are distinct fields, there are certain areas of intersection and possible synergy. Let's investigate their partnership in more detail.

1. Enhanced processing speed: Quantum computers may be able to carry out some operations significantly more quickly than conventional ones. Numerous AI tasks, including optimisation, machine learning, and data analysis, can benefit from this speedup. Grover's algorithm and quantum support vector machines are two examples of quantum algorithms that have the potential to increase computational efficiency for particular AI issues.
2. Quantum algorithms for machine learning: Traditional machine learning methods are intended to be improved using quantum computing resources. Pattern recognition, classification, and prediction tasks may be enhanced by QML techniques such as quantum neural networks and feature selection inspired by quantum mechanics.

3. Data analysis and optimisation: Using quantum computers, massive datasets may be processed quickly, and challenging optimisation tasks can be completed. Applications of AI that need to manage enormous volumes of data or optimise complex models, such recommendation systems, financial modelling, and supply chain optimisation, can profit from this capabilities.
4. AI algorithms influenced by quantum mechanics: Even in the absence of quantum hardware, researchers have looked into applying quantum-inspired AI techniques to enhance more traditional AI methods. These algorithms improve optimisation, sampling, and search techniques and can be used for AI tasks including reinforcement learning, computer vision, and natural language processing. They are inspired by quantum principles.
5. AI research using quantum computing: Research and development in AI can also make use of quantum computing. In order to better understand complicated AI models and algorithms and identify areas for improvement, quantum simulators can be used. Researchers can investigate unique AI systems and create new methodologies using quantum computing.
6. Challenges and restrictions: Practical quantum computers with suitable qubit counts and low error rates are not yet commonly available. Quantum computing is still in its early phases of development. Additionally, there are difficulties in creating quantum algorithms for AI tasks and integrating them with current AI frameworks. Due to the inherent probabilistic nature of quantum algorithms, it can also be difficult to comprehend and validate the results. It's crucial to highlight that much research is currently being done to determine how quantum computing and AI might complement one another, and that the full depth of this relationship is still being studied. As quantum technology develops, it has the ability to transform different facets of AI, opening up new possibilities and capabilities. Artificial intelligence (AI) and quantum computing are two quickly developing disciplines that have the potential to completely change a variety of industries. This study intends to investigate the connections between quantum computing and AI, stressing the potential advantages, difficulties, and prospective areas of cooperation between both fields. This examination provides light on the complex interactions between quantum computing and AI by reviewing the current research and developments as well as the future prospects.

6. ATOMIC ALGORITHMS MACHINE LEARNING REVOLUTION: FROM THEORY TO PRACTICE

Machine learning has made incredible strides in recent years, revolutionizing sectors, and changing how we approach difficult challenges. Among these developments, the idea of atomic algorithms has become recognized as a potent accelerator of machine learning development. Atomic algorithms are small, effective, and specialized algorithms created to handle particular jobs with great accuracy and efficiency. The goal of this paper is to study the potential of atomic algorithms in quickening the machine learning revolution and bridging the gap between theoretical advancements and real-world applications. It does this by delving into the theory and practice of atomic algorithms. The theoretical underpinnings of atomic algorithms are investigated, including their definition, properties, and guiding theories and approaches.

A full discussion of the benefits and difficulties of atomic algorithms is presented. The article discusses the advantages of atomic algorithms, including their improved performance for particular applications, increased computational efficiency, and decreased complexity. However, difficulties with generalizability, interpretability, and adaptability to other datasets are also discussed, highlighting areas in need of additional study and advancement. A variety of machine learning topics are described with real-world applications of atomic algorithms. These applications show in detail how atomic algorithms overcame particular difficulties and outperformed conventional methods in terms of performance. The study illustrates the transformational potential of atomic algorithms in real-world scenarios by looking at their impact on particular applications and sectors. Focusing on methods for creating and perfecting these algorithms, the exploration of atomic algorithm development and optimization is undertaken. To increase the efficiency and effectiveness of atomic algorithms, strategies for feature selection, algorithm specialization, and optimization are presented. Researchers and practitioners may now investigate and use these potent tools thanks to the overview of algorithmic frameworks and libraries facilitating the construction of atomic algorithms. One of the most important components of successfully implementing atomic algorithms is bridging the theory-practice divide. The use of atomic algorithms in real-world contexts is examined, as well as methods for converting theoretical developments into useful implementations. It is investigated how atomic algorithms can be adopted and integrated seamlessly with current machine learning workflows and frameworks.

The ramifications for society and ethics are rigorously analyzed, addressing issues with fairness and bias in the context of atomic algorithms. Atomic algorithms' societal effects on various stakeholders are taken into account, coupled with responsible development and deployment procedures to assure moral and just use. Future directions and difficulties in atomic algorithm research and development are covered in the paper's conclusion. In this dynamic sector, new developments and potential research directions are emphasized to encourage further study and cooperation. The possible implications of atomic algorithms for machine learning in the future are discussed, highlighting the necessity for ongoing developments to fully realize their disruptive potential. This paper as a whole provides a thorough examination of atomic algorithms within the framework of the machine learning revolution. It provides insights into how these small, focused algorithms can revolutionize numerous industries, advancing machine learning and influencing the development of intelligent solutions by bridging the theoretical and practical divide.

7. A QUANTUM LEAP FOR SMARTER AI SYSTEMS: QUANTUM MACHINE LEARNING

Artificial intelligence is advancing thanks to the confluence of quantum computing and AI, which has also revolutionized the capabilities of intelligent systems. This article focuses on the idea of QML, a cutting-edge strategy that uses quantum-computing concepts to improve conventional machine learning techniques. The article begins by exposing readers to the interesting nexus between quantum computing and artificial intelligence while highlighting QML's revolutionary potential. It draws attention to the importance of this developing topic and the advantages it might provide in terms of allowing smarter and more effective AI systems. The paper introduces key quantum computing ideas to create a strong basis. It defines qubits, the fundamental units of quantum systems, and explains important characteristics like superposition and entanglement. It also discusses fundamental quantum operations and quantum gates, which serve as the building blocks of quantum computers. To familiarize readers with the fundamentals of quantum computing, an overview of the quantum circuit model and quantum computational complexity is also presented. The article then digs into machine learning's foundations, introducing traditional machine learning algorithms and approaches. It covers the supervised learning, unsupervised learning, and reinforcement learning paradigms, which are the three primary types of machine learning. Additionally, the difficulties and restrictions of conventional machine learning are examined, laying the groundwork for the introduction of QML as a potential remedy.

In-depth research is done on the fundamental ideas behind quantum machine learning. The article demonstrates how quantum computers can be more advantageous for machine learning applications than classical computing. It focuses on how quantum interference and quantum parallelism might increase computer efficiency and make it possible for learning algorithms to become more potent. In order to show how quantum systems can offer new representations and transformations of data for machine learning, the concepts of quantum feature space and quantum state preparation are also covered. The article's conclusion discusses quantum machine learning's prospective uses and implications. In particular, it demonstrates how QML may improve a number of AI tasks, including pattern identification, optimization, and data analysis. To demonstrate the revolutionary power of QML in practical situations, examples of prospective use cases in sectors including healthcare, banking, and cybersecurity are given. This article provides a thorough examination of quantum machine learning, demonstrating its potential to significantly increase the power of AI systems. QML opens up new vistas for resolving complicated issues and achieving previously unheard-of levels of intelligence in AI systems by fusing the strength of quantum computing with the principles of machine learning.

8. AI QUANTUM OPTIMIZATION: THE FUTURE OF MACHINE LEARNING

Machine learning has advanced significantly in recent years, but new strategies are required to push the limits of AI further as complicated issues keep cropping up. AI Quantum Optimization is one strategy that uses the capabilities of quantum computing to transform machine learning. In this article, the idea of AI quantum optimization is examined, emphasizing its potential to shape the direction of machine learning. Setting the stage early on, the article emphasizes the necessity of advanced optimization methods in machine learning. It examines the difficulties encountered by conventional optimization algorithms when tackling complex, high-dimensional issues as well as the difficulties they have in quickly locating the best solutions. The next section of the paper covers the idea of AI Quantum Optimization, which makes use of quantum computing principles to improve machine learning optimization processes. It describes how quantum computers have the potential to accelerate optimization and find optimal solutions more quickly due to their capacity to analyses enormous volumes of information concurrently and explore several options at once. It then delves into ideas like quantum annealing, quantum gates, and quantum states to examine the theoretical underpinnings of AI quantum optimization. It demonstrates how these ideas can be used to address issues with parameter tuning, feature selection, and model optimization in machine learning.

The article also illustrates the potential applications of AI quantum optimization in different fields. It emphasizes how this method can, among other things, optimize neural network topologies, hyper parameter optimization, recommendation systems, and financial portfolio management. The revolutionary effects of AI Quantum Optimization are demonstrated through examples drawn from everyday life and success stories. The paper concludes by examining AI quantum optimization's potential and how it may affect machine learning in the future. It highlights current research initiatives, developments in quantum hardware, and the possibilities for hybrid strategies that mix traditional and cutting-edge quantum computing methods. The transformational potential of AI Quantum Optimization in enhancing the capacities of machine learning systems and guiding the future of AI is highlighted in the article's conclusion. In decision, this article offers a thorough examination of AI Quantum Optimization as the upcoming development in machine learning. It demonstrates the benefits of applying quantum-computing ideas to optimization problems and offers examples of possible uses for AI quantum optimization in a range of fields. AI Quantum Optimization lays the way for tackling difficult optimization problems and opening up new horizons in machine learning by fusing the strength of quantum computing with the intelligence of machine learning.

9. ENHANCED QUANTUM LEARNING DEFINING NEW AI BOUNDARIES

In recent years, artificial intelligence (AI) has made incredible strides, allowing machines to carry out difficult tasks and make wise decisions. Traditional AI algorithms still have some limitations, though, when it comes to handling some challenging issues. By fusing the capabilities of quantum computing with machine learning strategies, enhanced quantum learning is an exciting new frontier that aims to push the limits of artificial intelligence. This article examines the idea of enhanced quantum learning and how it can open up new AI horizons. The article starts out by outlining the current difficulties and constraints faced by conventional AI algorithms. It emphasizes the requirement for novel strategies that can get over these restrictions and more successfully deal with difficult issues. This prepares the ground for Enhanced Quantum

Learning to be discussed as a potential solution. The paper then explores the fundamental ideas behind machine learning and quantum computing. It gives a succinct explanation to quantum computing, outlining the potential for exponential computing power and defining essential concepts like superposition and entanglement. To give readers a thorough knowledge of conventional AI methods, it also covers the principles of machine learning, such as supervised and unsupervised learning, reinforcement learning, and deep learning.

The fundamental ideas behind Enhanced Quantum Learning are then investigated. The article describes how, by taking use of the characteristics of quantum systems, quantum computing might improve machine learning methods. It emphasizes how quantum entanglement, quantum interference, and quantum parallelism have the potential to make learning algorithms faster and more effective. It also examines how more accurate and reliable data representation and processing made possible by quantum computing can enhance AI task performance. The paper then goes on to highlight Enhanced Quantum Learning's potential uses in a variety of fields. It describes how this method can be used to address complex optimization issues, improve natural language processing, increase the accuracy of picture and speech recognition, and, among other things, optimize resource allocation. Examples from the real world and case studies are used to demonstrate how Enhanced Quantum Learning has completely changed the way that real-world AI applications operate. The article also examines the difficulties and factors to be taken into account while applying enhanced quantum learning. It discusses the current constraints of quantum hardware, the requirement for specialized knowledge in both machine learning and quantum computing, and the integration of quantum algorithms with current AI frameworks. It highlights how crucial interdisciplinary cooperation and ongoing study are to resolving these problems. The short piece ends by examining the potential applications of enhanced quantum learning and how they can progress artificial intelligence. It examines current research initiatives, developments in quantum hardware, and the possibilities for hybrid strategies that mix traditional and cutting-edge quantum computing methods. The articleemphasises how Enhanced Quantum Learning is pushing the frontiers of AI and opening up new avenues for the solution of challenging issues. In this regard, this study offers a thorough investigation of enhanced quantum learning as a potent strategy to push the limits of artificial intelligence. It emphasises how machine learning methods could be improved by quantum computing, offering more effective and precise answers to challenging issues. Enhanced Quantum Learning expands the boundaries of AI by fusing the strengths of quantum computing and machine learning, paving the path for previously unimaginable developments in intelligent systems and applications.

10. UNRAVEL NEW DIMENSIONS OF INTELLIGENCE WITH QUANTUM-ASSISTED MACHINE LEARNING

Numerous sectors have been altered by machine learning's quick advancement, but as long as difficult challenges exist, researchers are always looking for fresh ideas to expand the capabilities of AI. Quantum-Assisted Machine Learning (QAML), which combines the strength of quantum computing with conventional machine learning methods to uncover new dimensions of intelligence, is one such potential path. This article examines the idea of QAML and how it might revolutionize the field of artificial intelligence. The first section of the article provides readers with an overview of the state of machine learning and the difficulties that conventional algorithms have when attempting to solve complex issues. It highlights the requirement for cutting-edge computational tools capable of handling massive datasets and high-dimensional data representations. The article then delves into the fundamentals of quantum computation, giving a background knowledge of qubits, quantum gates, and quantum operations. It clarifies crucial quantum phenomena like superposition and entanglement, which allow quantum computers to process information in a fundamentally different manner from conventional computers.

The article introduces the idea of quantum-aided machine learning by building on this foundation. It discusses how quantum computing can be used to improve a number of machine learning processes, such as feature selection, model training, and data preprocessing. It describes how qubits can simultaneously encode and process information, potentially accelerating computation exponentially. The article then looks at the potential uses of QAML in several fields. It demonstrates how QAML may be applied to a variety of tasks, including complicated pattern recognition, large-scale system optimization, natural language processing, drug development, and financial modelling. Examples from the real world and case studies are used to illustrate how QAML has significantly changed the way that real-world AI applications are developed.

The article also covers the difficulties and factors to be taken into account when using QAML. It discusses the constraints of present quantum hardware, the requirement for specialized knowledge in both quantum computing and machine learning, and the incorporation of quantum algorithms with current machine learning frameworks. It emphasizes how critical it is to create scalable, reliable quantum algorithms and to maximize the fusion of classical and quantum systems. The future prospects of QAML and their consequences for the development of AI are covered in the article's conclusion. It examines current research initiatives, developments in quantum hardware, and the possibilities for hybrid strategies that mix traditional and cutting-edge quantum computing methods. To realize the full potential of QAML, the article emphasizes the necessity for ongoing interdisciplinary cooperation and financial support for quantum computing research. In decision, this study offers a thorough investigation of quantum-assisted machine learning (QAML) as a game-changing strategy to uncover fresh levels of intelligence in AI systems. It demonstrates the multiple ways in which machine learning can be improved by quantum computing and the wide range of disciplines in which QAML is being applied. QAML expands the boundaries of AI, allowing for breakthroughs in the solution of complicated problems and paving the way for the next generation of intelligent systems by fusing the strengths of quantum computing with machine learning.

11. UNLOCKING UNPRECEDENTED INSIGHTS AND CAPABILITIES WITH QUANTUM ALGORITHMS FOR AI

Artificial intelligence is not an exception when it comes to the potential influence of quantum computing, which has emerged as a groundbreaking technology. The idea of using quantum algorithms for AI to achieve previously unheard-of insights and capabilities is explored in this article. Researchers are able to get around conventional constraints and open up new opportunities for developing AI by utilizing the potential of quantum computing. The article starts off by giving a general summary of the current obstacles in AI, including managing intricate and sizable datasets, improving machine learning models, and resolving computationally demanding issues. It highlights the requirement for creative solutions to meet these difficulties and expand the capabilities of AI. The paper then presents the basics of quantum computing to the audience. It explains the concepts of superposition and entanglement, which allow for parallel computation and exponential computing capacity in quantum systems. In order to comprehend quantum algorithms, it is also important to understand key ideas like qubits, quantum gates, and the quantum circuit model. The article investigates the possibilities of quantum algorithms for AI by building on this foundation. It explores particular quantum algorithms that could revolutionize many parts of AI activities, such as quantum support vector machines, quantum neural networks, and quantum clustering algorithms. It shows how these algorithms take advantage of the special features of quantum systems to produce previously unheard-of insights and talents.

The article also emphasizes the potential uses of quantum algorithms in AI. It talks about how jobs like data analysis, pattern identification, optimization, and natural language processing can be improved by quantum algorithms. Examples from the real world and case studies are given to illustrate how quantum algorithms may boost AI applications. The article also discusses the difficulties and factors to be taken into account while applying quantum algorithms to AI. It talks about how to integrate quantum algorithms with conventional machine learning frameworks, the current constraints of quantum hardware, and the necessity for expertise in both quantum computing and AI. It highlights the significance of creating quantum error correction methods, improving quantum circuitry, and encouraging interaction between the communities of quantum computing and AI.

The article also discusses the ramifications for the development of AI systems as well as the chances for quantum algorithms in the future. It highlights current research initiatives, developments in quantum hardware, and the possibilities for hybrid strategies that mix traditional and cutting-edge quantum computing methods. The article's conclusion emphasizes the revolutionary potential of quantum algorithms in revealing hitherto unheard-of insights and capabilities, leading to substantial developments in the field of AI. In finalization, this article offers a thorough examination of how quantum algorithms for AI can uncover hitherto unheard-of insights and capabilities. It emphasizes how quantum computing has the potential to surpass preexisting constraints and revolutionize numerous elements of AI activities. Researchers can open new doors and achieve advances in data analysis, optimization, and pattern identification by utilizing the capabilities of quantum algorithms. To fully realize the potential of quantum algorithms for AI, the article emphasizes the significance of ongoing research, collaboration, and quantum computing breakthroughs.

12. ALGORITHMS FOR QUANTUM MACHINE LEARNING: LAYING THE FOUNDATION FOR THE QUANTUM AI REVOLUTION

It is quite likely that a revolution in intelligent systems will be sparked by the convergence of quantum computing with artificial intelligence (AI). The revolutionary discipline of quantum machine learning (QML), which combines the capabilities of quantum computing with the fundamentals of machine learning, has developed. In this article, the significance of algorithms in quantum machine learning and how they create the framework for the quantum AI revolution are examined. The importance of quantum machine learning in overcoming the drawbacks of traditional machine learning algorithms is covered in the article's opening paragraphs. It emphasizes the quantum computers' exponential processing capacity and how this might fundamentally alter the capabilities of AI systems. In order to comprehend quantum machine learning algorithms, the paper then digs into the fundamental ideas of quantum computing. It clarifies the fundamental concepts of superposition, entanglement, and quantum interference that serve as the foundation for quantum algorithms. Qubits and quantum gates are also introduced, emphasizing their significance in quantum calculations. On top of this basis, the article investigates various quantum machine learning methods. The article examines quantum variations of conventional machine learning methods, including quantum SVMs, quantum neural networks, and quantum clustering algorithms. It describes how these algorithms take advantage of quantum features to strengthen learning tasks, advance optimization procedures, and manage complex data representations. The article also emphasizes the significance of algorithmic creation and improvement in quantum machine learning. It examines difficulties in using quantum algorithms, including quantum error correction, circuit depth optimization, and utilization of quantum resources. It emphasizes the necessity of algorithmic improvements to fully utilize quantum computing's potential in AI. The article also looks into possible sectors where quantum machine learning techniques could be used. In areas including drug development, financial optimization, recommendation systems, and quantum chemistry simulations, it explains how quantum algorithms can aid in the solution of challenging issues. The revolutionary influence of quantum machine learning algorithms is demonstrated using case studies and real-world situations.

The article's conclusion discusses the potential applications and ramifications of quantum machine learning methods. It examines current research initiatives, improvements in quantum technology, and the possibility for hybrid strategies that blend conventional and novel quantum algorithms. The article highlights the crucial role of algorithms in creating the groundwork for the quantum AI revolution, advancing intelligent systems, and opening up new avenues for resolving challenging issues. In final analysis, this study offers a thorough investigation of quantum machine learning algorithms and their function in setting the groundwork for the quantum AI revolution. It examines numerous quantum machine-learning techniques and emphasizes how quantum computing might improve machine-learning problems. Researchers are laying the groundwork for revolutionary developments in AI systems by creating and improving these algorithms. In order to fully utilize the capabilities of quantum computing in AI, the article emphasizes the significance of ongoing research and collaboration in algorithm development.

SUMMARY

In order to improve AI systems, QML combines the strength of quantum computing with the concepts of machine learning. This article investigates QML's potential to improve AI and fuel revolutionary developments. The paper opens by pointing out the shortcomings of traditional machine learning algorithms when it comes to solving complicated issues. It introduces the idea of quantum computing and describes how special features like superposition and entanglement might fundamentally alter the power of AI systems. The paper then delves into the foundations of QML. It examines qubits, quantum gates, and quantum circuits as well as the theoretical underpinnings of quantum computing. In particular, it highlights how quantum parallelism and quantum interference could speed up machine learning procedures. This foundation is built upon as the article covers QML's uses in AI. Quantum iterations of well-known machine learning algorithms are covered, including quantum neural networks, quantum SVMs, and quantum clustering methods. It serves as an illustration of how QML can enhance processes like pattern recognition, optimization, and data analysis. The article also discusses the difficulties and factors to take into account when using QML. It talks about how to integrate QML algorithms with current AI frameworks, the current constraints of quantum hardware, and the demand for specialized knowledge in both quantum computing and machine learning. It emphasizes how critical it is to create scalable, reliable quantum algorithms and to maximize the fusion of classical and quantum systems. The article also addresses QML's potential for development and how it might advance AI. It examines current research initiatives, developments in quantum hardware, and the possibilities for hybrid strategies that mix traditional and cutting-edge quantum computing methods. The transformative potential of QML in enhancing AI system capabilities and opening up new horizons in the field of AI is highlighted in the article's conclusion. In conclusion, this article gives a general overview of quantum machine learning (QML) and how it can improve artificial intelligence (AI) using quantum computers. It draws attention to the potential for quantum phenomena like superposition and entanglement to fundamentally alter machine learning techniques. Researchers can develop AI fundamentally by utilizing QML algorithms, enhancing operations like pattern recognition, optimization, and data analysis. The article emphasizes the need for continued research, collaboration, and advancements in quantum computing to fully harness the potential of QML in enhancing AI.

Chapter 9
An Enhanced Study of Quantum Computing in the View of Machine Learning

Arvindhan Muthusamy
Galgotias University, India

M. Ramprasath
https://orcid.org/0000-0002-2667-9184
SRM Institute of Science and Technology, India

A. V. Kalpana
https://orcid.org/0000-0003-2289-4968
SRM Institute of Science and Technology, India

Nadana Ravishankar
https://orcid.org/0000-0002-9854-5150
SRM Institute of Science and Technology, India

ABSTRACT

Emerging technologies, including quantum information science and artificial education systems, have the potential to have significant implications for the future of human civilization. Quantum information, on the one hand, and machine learning (ML) and artificial intelligence (AI), on the other, consume their personal unique set of queries and contests that have been studied in isolation up until now. However, a recent study is starting to examine whether these disciplines can teach one another anything useful. The discipline of quantum ML investigates how quantum computing and ML may work together

DOI: 10.4018/978-1-6684-7535-5.ch009

Copyright © 2023, IGI Global. Copying or distributing in print or electronic forms without written permission of IGI Global is prohibited.

to find solutions to challenges in both areas. Major advancements in the two areas of effect have been made recently. Particularly relevant in today's "big data" era is the use of quantum computing to speed up the solution of machine learning (ML) challenges. However, ML is already present in many state-of-the-art technologies and may play a crucial role in future quantum technologies.

1. INTRODUCTION

Quantum theory has impacted the entirety of the physical sciences. Depending on the field, this impact can be subtle or dramatic, but it often manifests itself at the microscale. In the latter half of the twentieth century, engineers began taking advantage of true quantum effects; these effects offer characteristics superior to those possible using solely classical methods. Classical systems. When such engineering first began, such as the laser, transistors, and nuclear magnetic resonance devices. The 1980s were a pivotal decade for the growth of the second wave, which is still not entirely comprehensive. Quantitative research on the feasibility of implementing quantum impacts for a wide range of activities, which, at their core level, be concerned with information handling (Brookes, 2017). This includes science and technology fields like cryptography, computers, and sensing, and All branches of metrology now speak the same language because to The Study of Quantum Information. Studies investigating these topics programmes that brought together experts from different fields. Examples include the development and widespread use of quantum computing, communication, cryptography, and metrology. Fields of study have profoundly affected how we view information and its processing (Xia, 2018). An improved method for calculating molecular potential energy surfaces using a quantum algorithm that combines classical techniques with a constrained Boltzmann machine.

The smallest quantifiable unit of a physical phenomenon is called a "quantum" in physics. Quantum theory seeks to determine the likelihood of the presence of a quantum particle at a specific location (Wittek, 2014). Quantum Machine Learning (QML) has grown and changed in computer science over the past few decades since it is related to Machine Learning (ML), the study of how to process best and analyse data for actionable insights. Appropriate data management is essential as its volume grows by 20% annually (Shore, 1994; Biamonte, 2017). Especially compared to conventional computers, quantum computers may be capable of quicker machine learning, which has prompted the study about how to design and deploy quantum software for this purpose. Niemann (2016) has discussed how useful Logic Synthesis for quantum state generation is.

There is a subfield of computer science known as quantum computing, which applies the ideas of quantum physics to data processing and manipulation. Classical computers, often known as traditional computers, represent and process information using bits. Bits can be either 0s or 1s. In contrast, quantum computers' qubits (quantum bits) can be in both the 0 and 1 states at once (Biamonte, 2021). Despite tolerant of faults, quantum computers are intriguing. They are unlikely to become commercially available anytime soon. Limitations on the amount of qubits and the circuit depth available in today's quantum technology are severe issues. Jerbi et al. (2021) has used deep reinforcement learning concept to enhance the performance of quantum computing model. Wittek (2014) aims to provide a synthesis that describes the most crucial machine learning methods within a quantum context.

Saharia et al. (2019) proposed optimising a constrained multivariable function without programming. This idea has since formed the foundation of machine learning, which uses algorithms to generate decision functions by establishing a correspondence between input and output data. To categorise labelled data categories, the most popular supervised technique in QML is the Quantum Support Vector Machine (QSVM) (Saharia et al., 2019), which employs the vector space optimisation bound of the sophisticated measurement. QSVM can reduce and develop patterns from unlabeled data (Pittman et al., 2021). In contrast to the structured patterns generated by machine learning, quantum systems generate unstructured patterns and explore the development and implementation of quantum software to speed up machine learning (Shor, 1994). Kandala et al. (2017) show that it is possible to experimentally optimize Hamiltonian problems containing as many as six qubits and over a hundred Pauli terms, calculating the ground-state energy for particles of varying sizes all the way up to BeH2. Amin (2016) has used the quantum Boltzmann distribution of the transverse-field Ising Oscillator to present a novel methods for machine learning (Deutsch, 1985).

This is proven that quantum theory and the 'universal quantum computer' are consistent with the idea of compatibility, and a class of model computing machines that is the quantum generalization of the group of Turing computers is explained. In the 20th century, supercomputers were developed to perform various analyses on mathematical models. In the 1950s, Knill et al. (2001) constructed artificial neural networks; from the 1960s to the 19902, Koashi et al. (2001) advocated deep learning via the backpropagation approach.

1.1 Quantum Technology Is Based on the Following Principles of Quantum Mechanics

Superposition: The superposition of states is a property of qubits that allows them to represent many possibilities concurrently. For instance, a qubit's state can have varying probabilities, representing 0 and 1.

Intertwined: The state of one qubit can get entangled with another qubit, even if the two qubits are separated in space. Regardless of how far apart two entangled qubits may be, measuring the state of one of them immediately influences the state of the other.

Quantum interference: Measurement-based probabilities are affected by both positive and negative interference between superposed states of qubits, a phenomenon known as quantum interference.

Figure 1. Basics of qubits

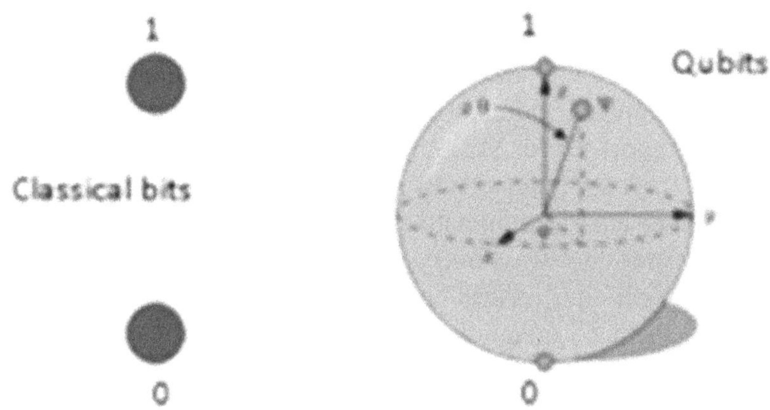

Because of these quantum effects, quantum computers can solve some computation tasks more quickly and accurately than classical ones. The use of quantum algorithms, which take advantage of superposition and entanglement to test several hypotheses simultaneously, may greatly accelerate the performance of some jobs by an exponential factor as shown in Figure 1. The factoring algorithm developed by Shor and the random-number-generating algorithm developed by Grover is two examples of well-known quantum algorithms. Grover (1986) examines multiple labels at once in quantum mechanical systems. It can be in a superposition of states. By timing operations correctly, successful analyses may bolster one another while unsuccessful ones interfere at random.

Superconducting circuits, trapped ions, topological qubits, and quantum dots are just a few physical systems used to construct quantum computers. The delicate quantum states in these systems are extremely vulnerable and must be protected at all costs. Maintaining coherence (the stability of qubits' quantum states), minimising errors due to noise, and increasing the number of qubits to execute sophisticated computations are all obstacles that must be overcome in developing quantum computers. Nielsen et al. (2002) present broad context making it accessible to anyone without specialist training in quantum mechanics or physics.

The factoring of big integers and the computation of discrete logarithms are two examples of complex algebraic equations that quantum computers are well-suited to solve (Sculd, 2015). These extraordinary skills stem from the qubits, are interpolating, juxtaposition, which-is, and interaction that are central to the theory of quantum mechanics and its unique principles and features. Improvements in encryption, Vast statistics study, QML, optimisation, the (IoT), and Blockchain are just a few of the many scientific fields that could benefit from quantum computing and as shown in Figure 2 (Arute et al., 2019; Biamonte et al., 2017).

1.2 Quantum Entanglement and Interference

The quantum states of several atoms become entangled, meaning that the state of any one component can't be explained in isolation from the states of the others. Quantum machine learning relies heavily on entanglement because it allows for the encoding and processing of intricate connections and relationships between quantum data. Quantum interference can effectively combine and magnify distinct possibilities in machine learning. Interactions between quantum states can be either constructive or destructive, depending on the nature of the quantum states involved. To increase their computational efficiency, quantum algorithms take advantage of this idea (Li et al., 2016). Graphical algorithms for learning the distribution of probability across a collection of inputs, newly presented as the fundamental components of multi-layer structures termed deep belief networks (DBNs), have received a lot of interest.

Figure 2. Integration of quantum computing and machine learning

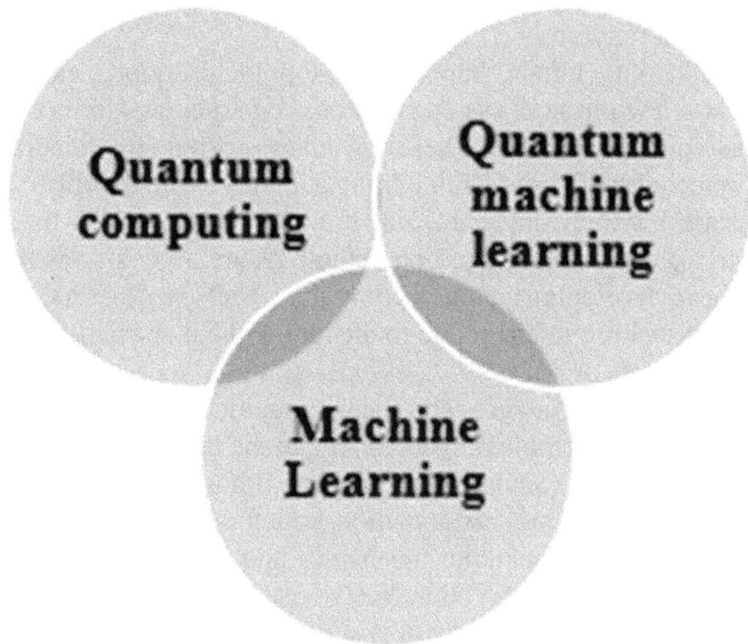

1.3 Quantum Measurement, Gates, and Operations

Interactions between quantum states can be either constructive or destructive, depending on the nature of the quantum states involved. To increase their computational efficiency, quantum algorithms take advantage of this idea. Quantum interference can effectively combine and magnify distinct possibilities in machine learning. The fundamental components of quantum journeys are quantum gates and operations. Quantum computing and state manipulation are made possible by gates like the controlled phase, CNOT, and Hadamard gates. Quantum machine learning algorithms use various quantum gates and operations to accomplish learning tasks and analyse quantum data.

The rest of the paper is organized as follow: section 2 present the detailed review on Quantum computing with respect to machine learning concepts.

2. LITERATURE REVIEW ON QUANTUM COMPUTING

In recent years, quantum machine learning has emerged as a hot topic in computer science. Despite its historical roots, modern studies have concentrated on several quantum machine learning methodologies and methods, such as supervised and unsupervised algorithms. Experiments on various datasets with complicated features are being conducted to merge and contrast the traditional and cutting-edge approaches to machine learning (Bausch, 2020). Presented RQNN combined with intensity amplification, the defined nano neurons that make up a QRNN cell provide a nonlinear activation of polynomials of its inputs and cell state, enabling the gathering of probability distributions across anticipated classes at each state. Mottonen et al. (2005) has discussed the setup of a computer equipped with quantum computing and certain quantum calculations make use of transformation.

Since the turn of the 20th century, scientists worldwide have tried to perfect and find uses for quantum computing. The quantum algorithm necessary for quantum computing was first introduced by Shor (1994; Saharia et al., 2019) in 1994. To encrypt messages or data conveyed through a channel utilizing a quantum Controlled-NOT gate, his proposal of a quantum method relied on the efficient factorization of big integers. The next step forward was shown when a quantum algorithm was demonstrated (Pittman et al., 2001). In the same year that Knill et al. (2001) and Koashi et al. (2001) established the concept of a polarisation beam splitter utilizing probabilistic manipulation of entangled photons, these authors came up with the probabilistic logic gates and their algorithm based on this idea. Givi et al. (2020) has discussed the role of quantum computing in aerospace engineering. Dong (2008) discussed the uses of reinforcement learning in quantum computing.

One significant theoretical work on quantum machine learning looks at the expense of computing and the necessity for transforming data, and it was conducted by Ciliberto et al. (2018). Quantum SVM, which provides quantum acceleration in machine learning applications, has been the subject of several studies (Aimeur, 2006). Datta et al. (2006) and Mttnen et al. (2005) examine the efficiency of machine learning algorithms run on classical and quantum computers, respectively, to illustrate the advantages and disadvantages of each. According to the latter, quantum multi-class SVM classifiers will be useful for high-qubit-count quantum computers of the future. Farhi (2014) has conveyed approximated answers to multimodal optimization issues using a quantum method. As p increases, so does the level of accuracy of the approximate produced by the algorithm. Vittorio Giovannetti (2008) demonstrates a design that cuts down the time it takes to make a memory call by a factor of ten or more. The number of switches quired drops from N in standard RAM designs (whether classical or quantum) to O(log N).

Support Vector Machines (SVM) and Quantum SVM (Shor, 1994) for data classification are most employed in both conventional and quantum machine learning. To solve classification problems, many classical and quantum SVM algorithms have been creating issues, as evidenced by a variety of benchmark analyses. For instance, Vittorio Giovannetti (2008) compares the performance of conventional and quantum SVM methods by implementing a quantum support vector machine (QSVM) on the MNIST corpus of handwritten images digits. The accuracy of QSVM and SVM are compared with Lloyd et al. (2013) and apply a classification model to a breast cancer image dataset. Lloyd proposes a QML-based recommendation system that can reach exponential improvement, whereas Durr and Hoyer (1996) use quantum computers that are prone to noise to test out quantum algorithms. Knill et al. (2001) has discussed different schema for quantum computation model. Sculd et al. (2017) discussed the implementation of distance-based classifier algorithm in quantum computing. Sculd et al. (2015) covered the basis introduction of quantum computing basic concepts.

2.1 Machine Learning Principles Generalised to the Quantum Universe

The advent of quantum theory required a new way of describing physical systems and a new definition of information. Quantum Information Processing QIP uses real quantum properties for faster processing (with quantum computers) and more efficient communication, making quantum information a universal concept. It is commonly argued that fundamental quantum features like the impossibility of exact duplication of even pure states lie at the heart of many quantum applications like cryptography. Comparatively, the evolution of a closed classical system corresponds to the (limited) group of permutations, whereas the evolution of a closed quantum system can go through any unitary evolution. Learning from and about data, or classical information, is the primary focus of most ML writings. The subject of how ML changes form as the data and Its processing quantum at its core may be. Quantum generalisations of supervised learning will be the first topic we cover, where the “points” in the data are now actual quantum states. This produces many cases that are equivalent in the standard setting (but not always the case; for instance, having one copy of an example and two copies of the same example are different). The next quantum generalisation of learning we will look at is the learning of quantum states, in which quantum states are utilised to represent the generalisations of unknown ideas in computational learning theory (COLT). Schuld (2019) investigate the theoretical underpinnings of this connection and demonstrate how it can inspire innovative approaches to the development of quantum machine learning algorithms. Shor (1994) has discussed Las Vegas methods which favor factor numbers and obtain separate logarithms on a quantum computer in an array of steps that is exponential in the size of the input.

2.2 Quantum Machine Learning (QML) Tasks

To create cutting-edge algorithms and models, researchers in Quantum Machine Learning (QML) have been combining ideas from quantum physics with traditional machine learning. It investigates how quantum technologies, such as quantum computers or simulators, might improve conventional machine-learning processes. Here are some important reminders and notes about QML projects:

To take advantage of quantum features like superposition and entanglement, QML explores the representation of classical data in a quantum manner. Information can be encoded using quantum states like qubits or quantum feature vectors. Quantum Neural Networks (QNNs) are developed with QML using quantum circuits or quantum-inspired models. To compute and learn, these networks make use of quantum gates and operations. QNNs may be more effective than traditional neural networks at resolving certain problems. Koashi et al. (2001) says quantum bits can be stored in the polarization of photons; therefore, we can use stochastic controleed-NOT and controlled-phase gates to access these bits.

QML investigates methods for selecting and extracting useful quantum characteristics from quantum data. Data dimensionality can be reduced through feature selection, while critical features can be extracted. Kopczyk (2018) has presented comprehensible and consistent explanations of quantum artificial intelligence techniques to a data scientist. The emphasis is not on formal mathematical presentation of the offered methods and formulas; rather, it is on providing clear, concrete examples and thorough, step-by-step explanations of the more challenging topics.

Quantum Machine Learning investigates quantum techniques for classifying data. Based on quantum measurements and quantum learning algorithms, quantum classifiers attempt to ascribe class labels to quantum data. These algorithms may offer benefits over traditional classifiers under certain conditions. Quantum Machine Learning investigates quantum algorithmic strategies for regression tasks. Predicting continuous variables using quantum data is the goal of quantum regression models. In some situations, quantum-inspired regression algorithms might be preferable to more traditional ones (Lloyd, 2013). Traditional techniques for this type of issue often have run times that are quadratic in the number of vectors and the dimension of the space being solved (Puskarov et al., 2020). Algorithms for machine learning frequently draw ideas from well-established conclusions and expertise in statistical physics.

QML explores quantum clustering methods by implementing quantum algorithms. The goal of quantum clustering is to collect comparable instances or patterns in quantum data. Compared to traditional clustering methods, algorithms influenced by quantum mechanics may reveal previously unnoticed patterns. QML investigates the creation of quantum generative models that can produce new samples of quantum data. To discover the underlying data distribution and produce new instances, these models use quantum states and operations. Data enhancement and synthetic data synthesis are two areas where quantum generative models may prove useful. QML explores how quantum systems can be incorporated into existing reinforcement learning architectures. Algorithms based on quantum reinforcement learning use quantum features to improve navigation, optimisation, and decision-making in uncertain settings. Liu (2022) has used the value function in quantum application and reinforcement-based quantum concepts which helps to solve multi-arm bandit problems. Wiebe et al. (2014) uses the deep learning-based algorithm for solving the conventical problem in quantum computing domain.

Quantum Machine Learning (QML) investigates quantum techniques for optimisation issues in conventional ML. To locate optimal solutions faster than traditional optimisation techniques, quantum optimisation techniques have been developed. These methods may be used for model optimisation, feature selection, and fine-tuning hyperparameters. Computing with QML often requires using either a quantum simulator or one of the soon-to-be-available quantum computers. Quantum computers can do quantum computations faster and more effectively, while quantum simulators allow researchers to replicate the behaviour of quantum systems. Pittman (2001) says with quantum logic with probabilities employing linear optical elements, some operations can be carried out, and some extra photons can be generated.

2.3 Learning From Quantum Data Using QLM

“Quantum generalisations,” also known as “quantum machine learning of quantum data,” describes using machine learning methods to examine and extrapolate from quantum data. Quantum data, as used here, is any data with a quantum component, such as a quantum state and size, or a quantum circuit. The purpose of quantum generalisations is to apply the concepts and characteristics of quantum mechanics to generate procedures and models that can analyse and extract insights from quantum data. These methods can help us better understand quantum systems, improve quantum simulations, and overcome obstacles in various quantum applications. Many essential features and methods of quantum generalisations in machine learning are as follows:

One of the primary obstacles in quantum generalisations is finding an appropriate illustration of quantum data for use in ML systems. Quantum data is often encoded using quantum states like qubits or quantum feature vectors. Both complicated vectors and dense matrices can be used to represent these states. The learning and processing of quantum data can be modelled after quantum circuits, which comprise quantum gates and operations. Learning algorithms for quantum circuits seek to optimise the parameters of these circuits so that they can carry out various tasks, including classification and regression, using the available quantum data.

Techniques for selecting and extracting features from quantum information are investigated through quantum generalisations analogous to conventional machine learning. The objective is to isolate the utmost significant quantum features that capture the essence of the data and eliminate the rest as noise. These methods lessen the data's dimensions and boost the effectiveness of further learning endeavours. Quantum generalisations include the creation of algorithms for classification and regression tasks that can be performed on quantum data. Both quantum regression models and quantum classifiers attempt to make predictions about continuous variables using quantum information. These algorithms improve the performance of their classical counterparts by using quantum phenomena like quantum interference and entanglement.

Clustering comparable instances or patterns in quantum data is also investigated by quantum generalisations, as are methods for reducing the dimensionality of quantum data. The quantum features used by quantum-inspired clustering algorithms make them effective in locating clusters. Also, the critical information in the quantum data can be preserved when employing dimensionality reduction techniques like quantum Principal Component Analysis (PCA). Quantum generative models aim to produce new quantum instances by learning the underlying distribution of quantum data. These models use quantum states and operations to signify the probabilistic behaviour of quantum systems. Applications of quantum generative models include the enhancement of data, the synthesis of synthetic data, and the simulation of complicated quantum systems as shown in Figure 3.

Quantum generalisations of reinforcement learning examine how quantum systems can be included in existing RL algorithms. Algorithms based on quantum reinforcement learning use quantum features to improve navigation, optimisation, and decision-making in uncertain settings. In some cases, these algorithms might even be superior to traditional forms of reinforcement learning. Quantum generalisations are also applied to optimisation issues in machine learning with the help of quantum algorithms. To locate optimal solutions faster than traditional optimisation techniques, quantum optimisation techniques have been developed. These methods can be used in various settings, including but not limited to hyperparameter tuning, model optimisation, and feature selection.

Figure 3. Classical digital circuit and quantum circuit

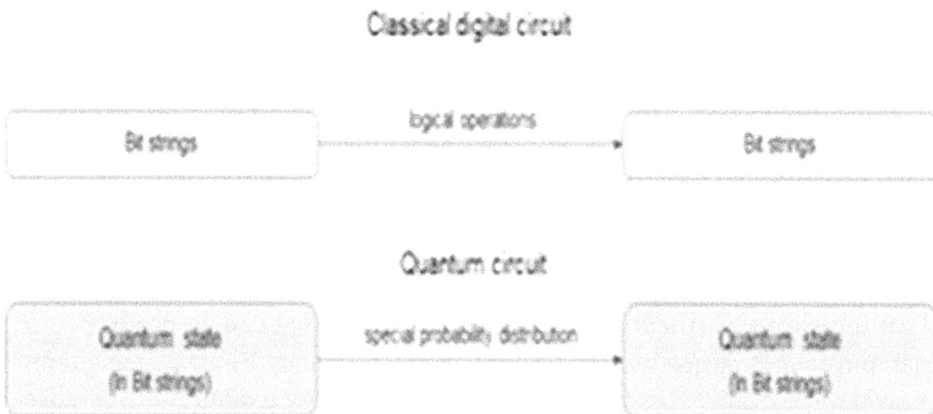

3. QUANTUM-ENHANCED MACHINE LEARNING

QEML is a method for improving the performance of supervised learning algorithms by employing quantum gates and orthogonal transformations. QEML aspires to answer the problems of data storage and delayed execution. Recognising the benefits of quantum supervised learning and the constraints of existing quantum algorithms is essential before attempting to build a Quantum K-Nearest Neighbours Algorithm. Improved representation space and accelerated algorithm execution due to the usage of quantum heuristics are two of QEML's major properties. Quantum superposition permits a reduction in storage requirements that is exponential.

One of the most important characteristics of quantum mechanics is superposition, which permits qubits to exist simultaneously in many states. Acceleration of machine learning algorithm implementation is quantum computing's second gift to the field of machine learning. QEML's "quantum parallelism" describes this characteristic. Since a quantum memory can be in a superposition state, "quantum parallelism" also results from this property.

It is possible to represent the individual parts of a superposition as functions. Each superposition part is graded according to its role in the quantum ledger.

3.1 Recent Developments in the Study of Machine Learning and Quantum Computing

Quantum machine learning algorithms translate traditional machine learning techniques into quantum algorithms. Amplitude amplification is the foundation of several quantum algorithms, including those for k-nearest neighbour and clustering. The technological routines for quantum matrix inversion or density matrix exponentiation are the backbone of quantum kernel approaches like support vector machines and Gaussian processes. Saharia et al. (2019) has discussed the way of solving the complex problem in quantum computing domain.

The current state of quantum computing can be visualised as a range of abstraction levels along which various paradigms have been developed. As a result, the learning curve for programmers and coders may increase. To approximate any arbitrary unitary operation, the most popular paradigm for quantum computing is the discrete variable gate model, in which a finite collection of unitary gates stands for logical transformations.

3.2 Regulating the Quantum World Using Reinforcement Learning

Researchers at Google suggest using quantum control with deep RL for general purposes, including quantum simulation, quantum chemistry, and quantum milestone tests. Creating a physical model for an operational quantum control process that accurately predicts error quantities is a major difficulty in modern quantum computing. The quantity of quantum information lost during the calculation, also known as "leakage," is significant since it not only causes errors that result in the loss of relevant quantum information but also gradually degrades the performance of a quantum computer. The Google team devised a quantum control cost function that considers leakage mistakes, control restrictions, total run-time, and gate infidelity to evaluate the accuracy of any released information. This allows the reinforcement learning methods to optimise these lenient punishment terms without sacrificing overall system control.

The novel quantum control cost function has inspired the creation of a powerful optimisation tool. They specifically chose an on-policy deep Reinforcement Learning approach called trusted-region reinforcement learning. On-policy Reinforcement Learning is useful in quantum systems because it can use non-local features in control trajectories, even if the control landscape is often high-dimensional and packed with many non-global solutions. The approach was effective across a wide range of test problems and was resistant to the effects of noise in the samples. Google thinks deep neural network reinforcement learning approaches have much room to grow in qubit control optimisation. Researchers have been inspired to embrace control approaches built on deep reinforcement learning due to their ability to harness non-local regularities of noisy control trajectories and to ease transfer learning between tasks.

3.3 Rise of Quantum Computing in Machine Learning

Quantum machine learning (QML) is a growing field of study that combines concepts from QC and ML. Using QML, users can integrate existing hardware into a single machine to create a quantum computer. As a result, it facilitated inefficient data manipulations using quantum machines founded on quantum theory, a very subtle body of physics law. It pioneers and entangles quantum computing in ways distinct from other fields' manipulation processes (Dinaburg et al., 2008). Unlike classical computers, quantum computers are predicted to be able to manipulate data with ease. Increased predicational output can be attributed to the QML procedures replacing the ML system, as seen in Figure 4.

Figure 4. Emergence of classical machine learning to quantum machine learning

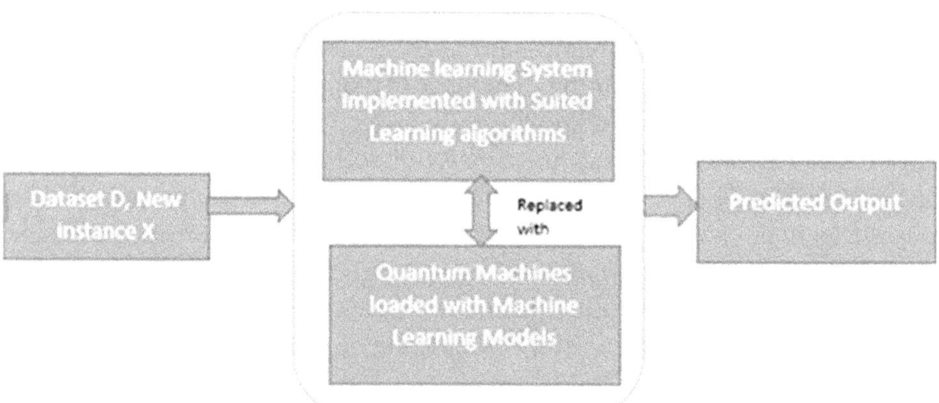

The building blocks of a quantum computer are quantum bits, or qubits, which use Boolean logic to output 0 and 1 alternately. The output of the quantum computer is verified by checking both the number of qubits and the related functions, which is sufficient for most applications. In QML, quantum computers modify the algorithms for learning to work with the specified problems. Variational algorithms, the building blocks of quantum machine learning, can only be run on specific kinds of quantum computers. The efficiency and structure of the learning algorithm change depending on the user. Different algorithms, such as Grover's and Shor's, are used depending on the number of qubits. However, only a few benchmarks are currently based on running ML algorithms on actual quantum hardware.

3.4 Usage of Quantum Computers

In cryptography, certain algorithms, such as those relying on factoring huge numbers, may be vulnerable to quantum computers. However, cryptographic approaches resistant to quantum computing are being developed to address this issue. Quantum computers may be used for optimisation, which has several practical uses, including logistics, supply chain management, portfolio optimisation, and resource allocation. Quantum computers are superior to conventional computers in their ability to simulate and model quantum systems. Materials research, pharmaceuticals, and quantum chemistry can all benefit from this.

Quantum computing can improve machine learning algorithms by either increasing the effectiveness of optimisation procedures or executing algorithms influenced by quantum mechanics. Research and development in quantum computing is lively and dynamic. While current quantum computers are still in their infancy and limited in qubit count and error rates, continual improvements hope to overcome these challenges and uncover the potential of quantum computing for tackling complex issues and revolutionising a wide range of sectors.

4. QUANTUM BASED MACHINE LEARNING ALGORITHMS

4.1 Support Vector Machine

(SVM) is a well-known Machine Learning technique employed for label classification of data. Originally developed in the 1960s, SVM aims to identify the hyperplane that maximizes the separation between labels in the dataset, commonly referred to as the “street” width. The provided figure illustrates this scenario in the case where the data has two features. In situations where the data possesses three features, SVM endeavours to locate the optimal plane, while for datasets with four or more features, it seeks to find the most suitable hyperplane as shown in Figure 5.

Quantum Support Vector Machines (QSVM) are an extension of classical Support Vector Machines (SVM) that leverage the principles of quantum computing to potentially enhance the performance of SVMs. While classical SVMs are widely used for classification and regression tasks, QSVMs aim to leverage the computational advantages of quantum computing to solve these problems more efficiently.

Figure 5. Classical support vector machine
Source: Wikipedia

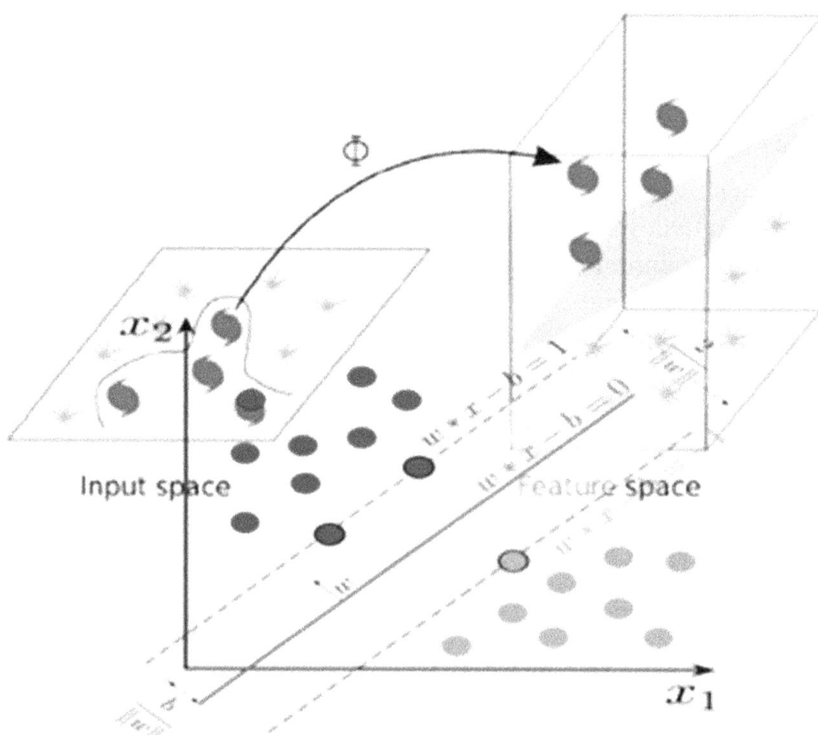

To understand QSVMs, let's first review the basic concepts of classical SVMs. A classical SVM is a supervised learning algorithm used for binary classification. It aims to find an optimal hyperplane in the feature space that maximises the gap between information points in different categories. As can be seen in the illustration, the decision boundary relies heavily on the support vectors (data points nearest to the hyperplane). The data points closest to the hyperplane, called support vectors, play a crucial role in defining the decision boundary as shown in the Figure 6. In contrast to the first method mentioned, the second method involves calculating the feature space within quantum circuits.

Figure 6. Mapping of SVM

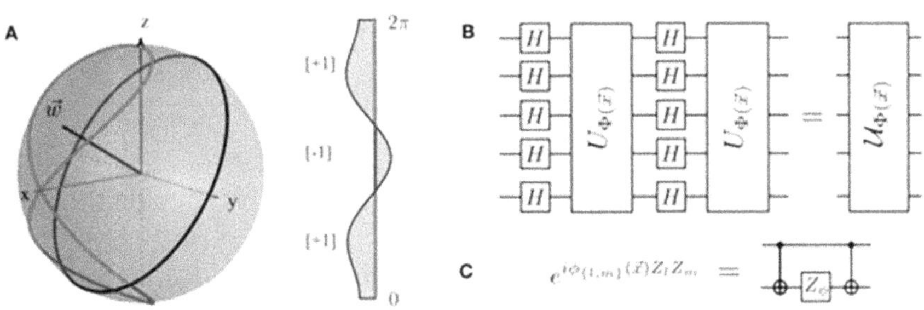

Now, let's delve into the quantum aspects of QSVMs. QSVMs rely on quantum algorithms, such as quantum phase approximation and quantum matrix transposal, to improve the efficiency of SVM training and inference. These algorithms take advantage of quantum parallelism and quantum interference to potentially perform certain computations more efficiently than classical counterparts.

Both methods share the use of quantum calculations to obtain the kernel function, departing from the traditional approach. However, to enhance the success rate, the data labels need to be provided in a superposition state. Although this requires reading the data from a quantum circuit, the opportunity to employ this data provision method is not available in the current context. Nonetheless, experimental studies have demonstrated that this method can achieve 100% predictive success, albeit with the utilization of specifically tailored data that can achieve such high accuracy.

4.2 Basics

In quantum computing, qubits serve as the fundamental units of computation. They can be in a superposition state, simultaneously representing both the $|0\rangle$ and $|1\rangle$ states.

A single qubit state can be expressed as a normalized two-dimensional complex vector, denoted as

$$|\psi\rangle = \alpha|0\rangle + \beta|1\rangle, \tag{1}$$

where $|\alpha|^2$ and $|\beta|^2$ represent the probabilities of observing the qubit in the $|0\rangle$ and $|1\rangle$ states, respectively.

This representation can also be visualized using polar coordinates θ and φ, where

$$|\psi\rangle = \cos(\theta/2)|0\rangle + e^\wedge(i\varphi)\sin(\theta/2)|1\rangle. \tag{2}$$

Here, θ ranges from 0 to π, and φ ranges from 0 to π.

The Bloch sphere is a representation of a single qubit state on the surface of a three-dimensional unit sphere. The tensor product of n single qubits, occurring in a superposition of 2^n base states spanning from $|00...00\rangle$ to $|11...11\rangle$, can be used to describe a multi-qubit system. A connection between various qubits in this system leads to the emergence of quantum entanglement. For instance, the observation of the first qubit immediately affects the state of the second qubit in a two-qubit system like Ö(1/2)$|00\rangle$ + Ö(1/2)$|11\rangle$. To perform quantum computations on these systems for predetermined goals, quantum gates in a quantum circuit are used.

4.3 Quantum Gates

Every quantum gate can be factorized into a combination of various fundamental operators, such as rotation operator gates and CX gates, as is known in classical computing. Quantum gates are unitary operators that transform one qubit system into another (Choi et al., 2020). A qubit state in a Bloch sphere is rotated by the rotation operator gates Rx(θ), Ry(θ), Rz(θ) around the corresponding axis by, and the CX gate entangles two qubits by flipping one of their states if the other is $\langle 1|$. It is commonly known that quantum algorithms can achieve exponential computational gains over existing algorithms in specific applications like prime factorization. These quantum gates use quantum superposition and entanglement to gain an edge over classical computing.

A kernel is a transformation function that takes data points and converts them into a different domain, making it easier to classify the dataset as shown in Figure 7. This process can involve adding new dimensions to the data or redistributing it within the existing domain, as shown in Figure 8. While computing the mapping itself can be computationally intensive, Support Vector Machines (SVMs) do not require the explicit representation of the map. Instead, SVMs rely on the dot products between the feature vector and the weight vector in the kernel function, as specified by the optimization equation. This feature makes kernels highly valuable in effectively classifying intricate datasets.

Figure 7. Kernel mapping

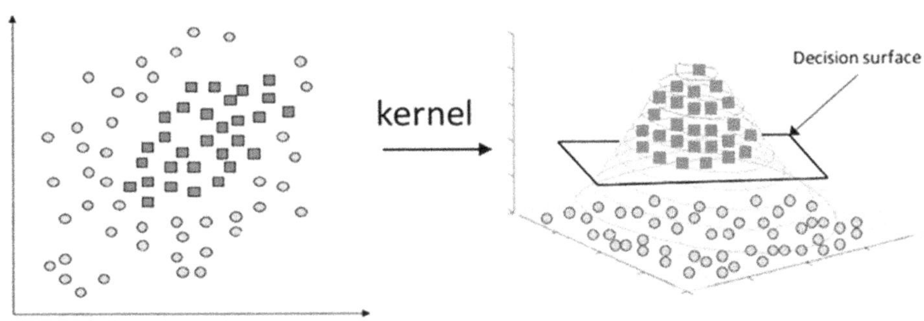

Figure 8. Example of redistributing data in the given domain
Source: Wikipedia

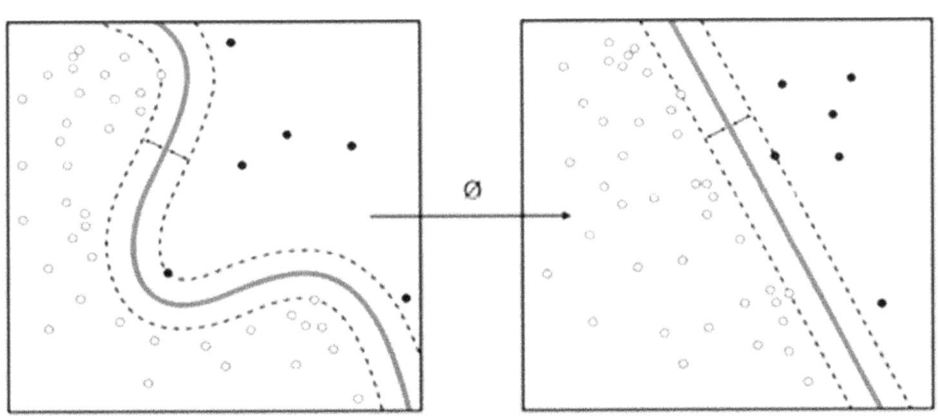

4.4 Quantum Kernel

Quantum kernels aim to influence the principles of QLM to enhance the process of mapping feature vectors and achieve improved outcomes. The underlying concept involves encoding classical data into qubits, applying quantum operations like superposition and rotations in the Bloch sphere, and subsequently computing the dot product of the resultant quantum states. These guidelines enable the utilization of quantum mechanics to enhance the mapping procedure. Several suggestions have been put forward for the kernel design. As an example, Aram Harrow et al. introduced the following circuit as a proposed kernel as shown in Figure 9:

Figure 9. Quantum kernel
Courtesy: Medium

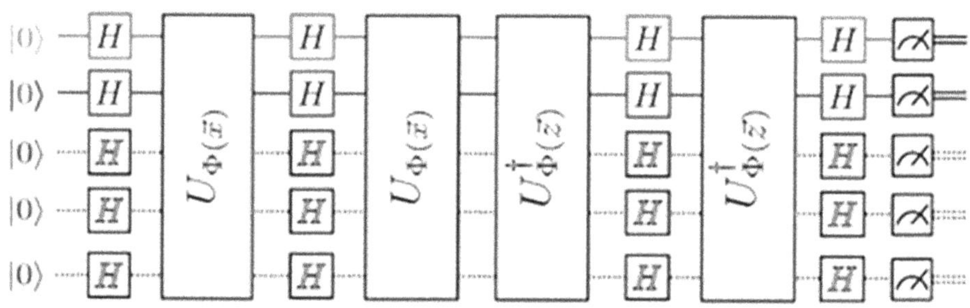

In this context, the unitary gate U is employed as a parameterized gate that operates on the feature vector. It represents the transformation described below, where S denotes the test dataset:

Quantum kernels must utilize operations that are beyond the capabilities of classical machines in order to surpass their performance, despite the inherent complexity involved. In the aforementioned case, the circuit incorporates Hadamard gates and Z Pauli matrices to gain an edge over classical machines.

4.5 ZFeature Map

The concept behind the ZFeatureMap as shown in Figure 10 is to assign a qubit to each dimension in the feature set. Each qubit undergoes a Hadamard gate operation followed by a unitary gate Uϕ.

Figure 10. ZFeatureMap circuit
Courtesy: Medium

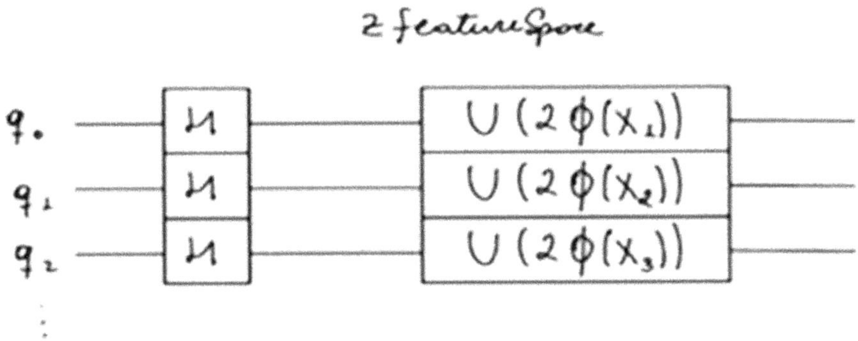

The circuit depicted above follows a linear structure, meaning that it does not transfer information across qubits using CNOT gates. Consequently, besides the H gates, this circuit does not exhibit any entanglement.

4.6 ZZFeature Map

On the other hand, the ZZFeatureMap as shown in Figure 11 introduces entanglement to the quantum system through the following approach.

Figure 11. ZZFeatureMap circuit
Source: Medium

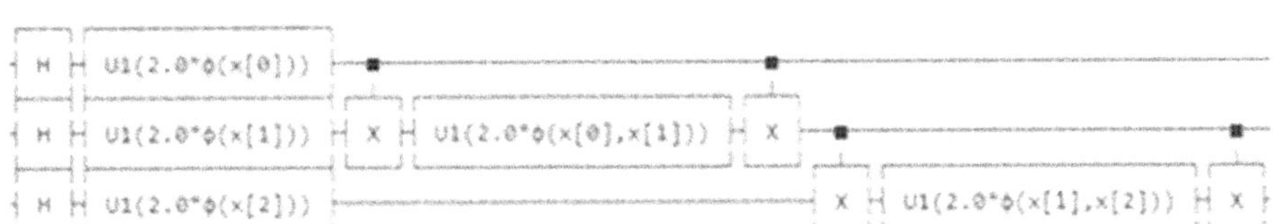

4.7 PauliMatrices Map

The concept behind the PauliMatricesMap as shown in Figure 12 involves incorporating Pauli matrices (X, Y, or Z) into the system to induce rotations of a specific magnitude on the Bloch Sphere.

Figure 12. PauliMatrices circuit with a Z, Y, and ZZ matrices
Source: Medium

5. QUANTUM NEURAL NETWORKS (QNN)

QNN are computer representations of neural networks that function according to quantum mechanical principles. Subhash Kak and Ron Chrisley first suggested the idea of quantum neural computation in 1995 in association with the notion of quantum mind, which contends that quantum effects have an impact on cognitive function. To create more effective algorithms, most research on QNN combines classical artificial NN models—which are frequently employed in machine learning for tasks like pattern recognition—with the advantages of quantum information. The difficulties in training classical neural networks, particularly in large-scale data applications, are what spurred this line of research. It is hoped that utilizing quantum computing characteristics like quantum parallelism, interference, and entanglement can produce beneficial resources. Since the development of functional quantum computers is still in its infancy, most quantum neural network models are currently only available as theoretical concepts awaiting physical experimentation.

Like their classical counterparts, the majority of QNN are built as FNN feed-forward networks. This arrangement transfers input from one layer of qubits to the following layer of qubits. Up until the final layer of qubits, each layer evaluates the data and passes the output on to the one above it. Contrary to classical neural networks, quantum neural networks' layers can vary in breadth, which means they may not always have the same amount of qubits as layers above or below them. Similar to traditional artificial neural networks, this structure's training entails figuring out the best routes. There are three main kinds of quantum neural networks: those that use traditional information on quantum computers, those that use advanced information on traditional computers, and those that use quantum data on quantum machines.

6. VARIATIONAL QUANTUM CIRCUITS (VQC)

Approximation, optimisation, and classification are only some of the numerical operations that may be performed by a variational quantum circuit (VQC) using gates that rotate operators with unbounded parameters. Using a classical computer for parameter optimisation makes the process known as a variational quantum algorithm (VQA) a combination of classical and quantum computing.

Many methods engaging VQC (Cerezo et al., 2020) are created to address different numerical problems (Choi et al., 2020) due to its universal function approximating property (Biamonte, 2021). This movement resulted in numerous VQA applications in machine learning and also allowed for the substitution of VQC for the artificial neural network in the preexisting model (Bausch, 2020; Cong et al., 2019; Dong et al., 2008; Schuld & Killoran, 2019). Despite sharing some similarities with artificial neural networks in that both rely on parameter learning for functional projections, VQC is distinct from these systems due to the nature of its computations. Since all quantum gate operations are reversible linear processes, multilayer frameworks in quantum electronics are created with entangled layers rather than activation functions.

6.1 Working of Quantum Neural Network

In this part, we attempt to illustrate the operation of a fundamental quantum neural network (QNN) using a straightforward example shown in Figure 13. A QNN processes data in the manner described below.

Figure 13. An example of a QNN using the parameters and linear entanglement as well as the input |ψ>

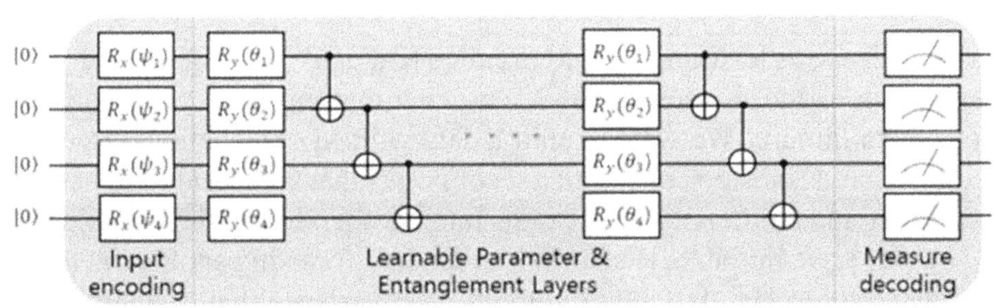

The input data is first encoded into the appropriate number of qubits' corresponding qubit state as shown in Figure 14. The qubit state is then altered for a certain number of layers using parameterized rotation gates and entangling gates. By determining the predicted value of a hamiltonian operator, such as Pauli gates, the converted qubit state is then measured. These measurements are recoded to create the necessary output data.

Figure 14. Illustration of a QCNN using a single convolution and pooling layer with the input $|i\rangle$, *the parameter*

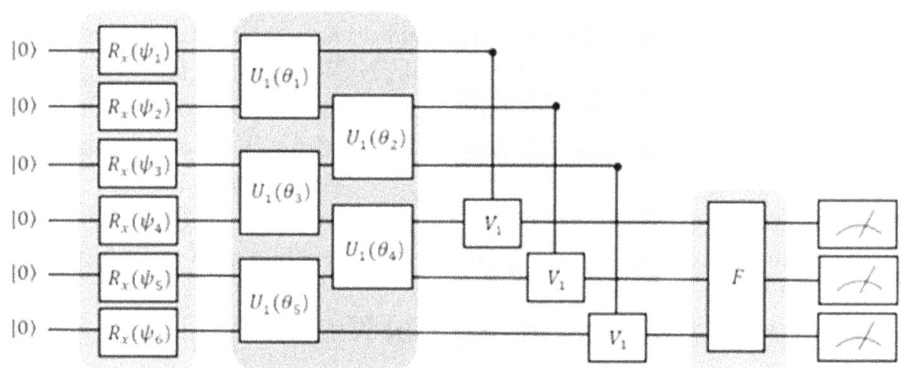

An optimizer like Adam optimizer will then adjust the parameters. It will be investigated how different functions can be played by a VQC-built neural network in the context of quantum neural networks.

7. QUANTUM K-MEANS CLUSTERING

An unsupervised clustering algorithm called K-means (Grover, 1996) divides n explanations into K clusters while ensuring that intra cluster variation is kept to a minimum. We want to split a data with M explanations, D = {a_1, a_2,... a_M}, into K clusters, or assign a set of points that are closer together into one cluster and a different set of points into another. Using the formula C = {c_1, c_2,... c_K}, we initialize K centroids at random (or using another method). Finding sets S = {Is_1, Is_2,... Is_K} with intra set variance that is smallest and inter set variance that is largest is the goal. Clusters are the collective noun for these sets.

The algorithm's mathematical aim is to discover:

$$\arg\min \sum_{i=1}^{k} \sum_{x \in S_i} \left\| x - c_i \right\|^2 \tag{3}$$

By determining the mean of each cluster, centroids are given new values after each iteration of the algorithm:

$$c_i = \frac{1}{L_i} \sum_{n=1}^{L_i} (x \in S_i) \tag{4}$$

where L_i denotes the set's size S_i. The procedure is then run once more after assigning fresh centroids to each cluster. You keep doing this until the algorithm converges. When clusters do not change any more, convergence has occurred. The number of features in the input vectors N, the total number of input vectors M, and K (number of clusters) all affect how time-consuming the classical version of the technique is. O(MNK)

For very high dimensional input vectors, the quantum version of the K-means algorithm offers an exponential speedup. The speedup results from the fact that the amplitude encoding only needs log N qubits to load N-dimensional input vectors. The quantum K-means method is presented in one variant (Kopczyk, 2018). The K-means clustering is carried out using three different quantum subroutines: SwapTest, DistCalc, and Grovers Optimization.

Swap Test

This test subroutine, which was initially used in (Aımeur et al., 2006), calculates the likelihood of measuring the control qubit in state $|0\rangle$ to determine the overlap between two quantum states $\langle a|b\rangle$. The overlap serves as a gauge of how similar two states are. States $|a\rangle$ and $|b\rangle$ are orthogonal if the control qubit's probability of being in state $|0\rangle$ is 0.5, but if it is 1, the states are identical. Only a measurement on the control qubit is needed, and states $|a\rangle$ and $|b\rangle$ might be unknown before the routine is conducted. Given is the probability that the control qubit is in the state $|0\rangle$

$$P(|0\rangle) = \frac{1}{2} + \frac{1}{2}|\langle a \mid b\rangle|^2 \quad (5)$$

Each of the states $|a\rangle$ and $|b\rangle$ has n qubits. It is also possible to load them from QRAM (Quantum Random Access Memory) straight away. (Giovannetti et al., 2008) or they can be generated using amplitude encoding (Mottonen, 2005; Niemann & Wille, 2016). After then, the overlap is determined by measuring the control qubit. Figure 15 shows the working principle of swap test circuit used in qubit state.

Figure 15. Swap test circuit

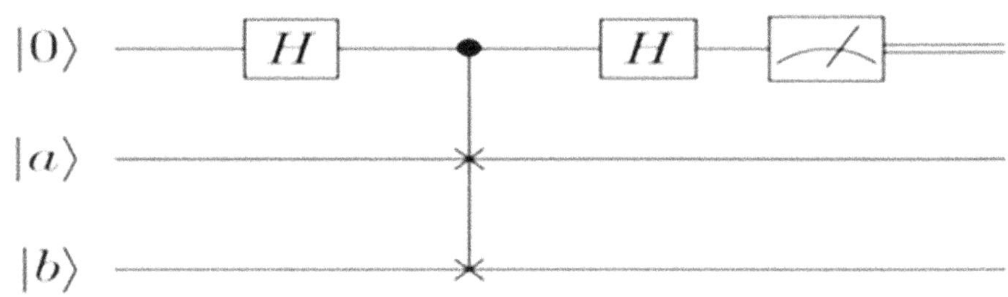

The DistCal function calculates the distance based on the Overlap from the SwapTest subroutine. The algorithm for utilizing SwapTest to calculate distance is outlined. The three procedures below are used to get the Euclidean distance $|a-b|^2$ between two vectors, a and b.

1) State Preparation: Two quantum states $|\psi\rangle$ and $|\phi\rangle$ are prepared as shown below:

$$|\psi\rangle = \frac{1}{\sqrt{2}}\left(|0, a\rangle + |1, b\rangle\right) \quad (6)$$

$$|\varphi\rangle = \frac{1}{\sqrt{Z}}\left(|a \mid\mid 0\rangle + |b \mid\mid 1\rangle\right) \quad (7)$$

Where $Z= |a|^2+|b|^2$

2) Find Overlap: Using SwapTest, the overlap $\langle\psi|\phi\rangle$ is calculated
3) Calculate Distance: The Euclidean distance is obtained using the following equation:

Distance $=2Z|\langle\psi|\phi\rangle|2$ (8)

K distances to each cluster centroid are computed using this method. The Grovers Optimization (Durr & Hoyer, 1996) subroutine, which is based on the Grovers method (Grover, 1996), is invoked to locate the closest cluster centroid. Then, the supplied point of information is placed in a group that includes the centroid's immediate neighbours. All the input data points are treated in this way. Centroids are recomputed by calculating the mean of all the data points in each cluster after each data point has been assigned a cluster. This step is performed till there is no anymore a need to reassign items to clusters.

The time-consuming aspect of the traditional K-means approach is determining the distance between N-dimensional vectors. Q K-means achieves an exponential speedup as a result of the effective quantum parallelism method for computing distances. The quantum version of the Lloyd's K-means algorithm (Grover, 1996) has a temporal complexity of O(log(N)MK), whereas the classical version has O(NMK). This makes sense when you realize that $\log_2 N$ qubits are used to encode the N-dimensional classical information.

8. QUANTUM REINFORCEMENT LEARNING

Quantum Reinforcement learning (QRL) has many potential to solve the problems in intelligent business decision-making, e.g., recommendation system, intelligent control for robot, self-driving, automation, electric vehicles charging scheduling and distributed optimization (Dong, 2008). Like classical reinforcement learning, QRL system can also be identified for three main sub elements: a policy, a reward function and a model of the environment (maybe not explicit). However, quantum reinforcement learning is differs in many ways from classical reinforcement learning and still has some problems, such as low learning efficiency and difficult to build model environment, which prevents the QRL algorithms from being directly applied in large-scale problems. Numerous approaches, such as expanding databases and collecting data in parallel, have been proposed to address QRL's issues. We can classify the four types of quantum algorithms for reinforcement learning (Liu et al., 2022) as follows: traditional algorithm paired with traditional data

(C-C), classical algorithm together with quantum information (C-Q), period algorithm together with conventional data (Q-C), and quantum properties. Method merged with quantum data (Q-Q). One type of C-C model is the classical RL model, which employs classical algorithms to address quantum physics challenges. The Q-C model describes a type of hybrid qu-classical RL, also known as quantum-enhanced RL, in which the value function is classical, but the procedure for iteration is quantum. In the Q-Q model, the value function and the algorithm are quantum-encoded.

In this section, we focus on designing the process of solving the problems based on Q-Q model and the process is given in the Figure 16.

Figure 16. The process of quantum-based reinforcement learning

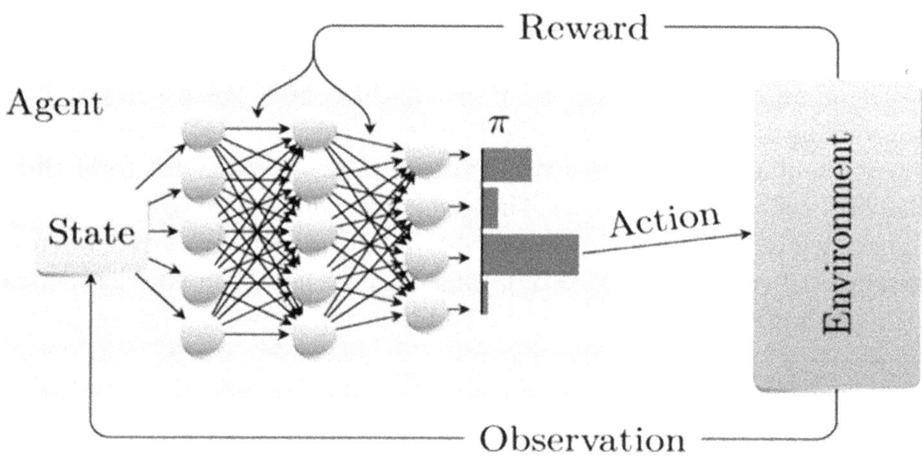

The QRL agent interacts with the external environment or scenario which could be a game, a simulated environment, or a real-world system and make decisions based on the policy and learned values. The state (quantum) represents the encoded information in the form of quantum bits (qubits). The measurement in the quantum learning collapses the quantum state, resulting in a classical outcome or observation based on the probability amplitudes of the quantum state. The measurement outcomes are used to compute rewards or feedback, indicating the performance of the agent in the environment. These rewards are crucial for updating the agent's policy and improving its decision-making. The observation/feedback obtained from the environment is used to update the QRL agent's policy and Q-values. This process involves adjusting the agent's strategies and reinforcing actions that lead to positive rewards while discouraging those with negative rewards.

A motivating application is to extend the QRL technique to environments of different sizes (Niels, 2023). With QRL, we can also consider the effect of complicating the environment by enlarging it. A second application of QRL is productive in considering the effect of taking multiple sequential steps. A third interesting application for QRL would be multi-agent learning under stochastic actions and with direct and quantum enhanced learning.

9. QUANTUM GIBBS BASED MACHINE LEARNING

Quantum Machine Learning (QML) is hybridization of classical machine learning techniques with quantum computation—is emerging as a powerful approach allowing quantum speed-ups and improving classical machine learning algorithms.

The quantum computing is capable of reducing the time required to train a restricted Boltzmann machine (RBM). The regular RBM models the probability of a given configuration of visible and hidden units by the Gibbs distribution with interactions restricted between different layers. Moreover, Gibbs state is the state of maximum entropy with the right expectation values.

Gibbs distributions have central role in Markov Chain Monte Carlo methods (E.g. estimating volume convex body), Machine learning (E.g. Boltzmann machines), and Optimization (E.g. matrix multiplicative updateHere, we delve into the generalised Gibbs ensemble (GGE), enriched by a handful of effective temperatures. These GGEs can serve as the foundation for a learning method analogous to a Boltzmann machine, which functions by discovering the best possible values for effective temps.

Boltzmann machines employ several hidden variables to learn the probability distribution of data, making them productive models. The variables of a standard Boltzmann machine are typically split into an "outer" or "visible" layer and an "inner" or "hidden" layer (Puskarov & Cubero, 2020). Since fully connected Boltzmann machines are unattainable, it is common practice to limit the relationships between the variables; for example, in a limited Boltzmann technology, the masses connecting the apparent and hidden layers are the only ones that do not vanish, while the weights within each layer are all zero. The Boltzmann distribution finds the probabilities $p(v^i; h)$ of configurations $(v^i; h)$ can be computed, as is done in statistical physics, from a, given by

$$p(v^i) = \sum_h \frac{e^{-\beta E(v^i,h)}}{Z}, \quad Z = \sum_{v^i,h} e^{-\beta E(v^i,h)} \tag{9}$$

Where v^i is the visible units which are fed the input data i and h are the hidden units. The parameter β corresponds to the inverse temperature and is traditionally set to β = 1, and $E(v^i; h)$ can be thought of as the energy of a given configuration.

In quantum Boltzmann machine learning, se the energy function is promoted to a Hamiltonian operator and the data is represented as a particular quantum state. A quantum state of Hamiltonian H, at temperature T is given by

$$\rho r = \frac{e^{-H/T}}{te(e^{-H/T})} \tag{10}$$

$$\infty \underset{\text{easy}}{\text{———}} \overset{T_c}{|} \underset{\text{hard}}{\text{———}} 0 \tag{11}$$

One drawback of quantum Boltzmann machines is that they are inefficient for large systems since their results can't be properly calculated through sampling (Mohammad, 2018). One strategy for learning the quantum algorithm involves setting a maximum value for the cross-entropy rather than searching for its minimum value. Minimising the upper limit requires estimating a quantity that can be done cheaply by sample. This bound-based strategy was demonstrated to be effective for several elementary datasets. Figure 17 (Xia, 2018) shows a proposed alternative method for training quantum Boltzmann machines, called state-based training, for data sets that may be described as a quantum density matrix. This method also enables the underlying source of the data used for instruction to be fundamental.

Assume a customizable quantum device, wherein the device's settings can be adjusted to modify the parameters of a calculation that would otherwise remain unmodifiable. We give some of these parameter names that correspond to the values of the input data x, and we link other parameters with variables that can be trained. The gadget gives us outputs y = f(x) that are sensitive to the values of x and the parameters we supply. This type of quantum device is related to variational (i.e. trainable) quantum circuits through a supervised learning model. Uncontrolled models can be built similarly (Li et al., 2016; Mishra et al., 2022).

Figure 17. Quantum Boltzmann machine

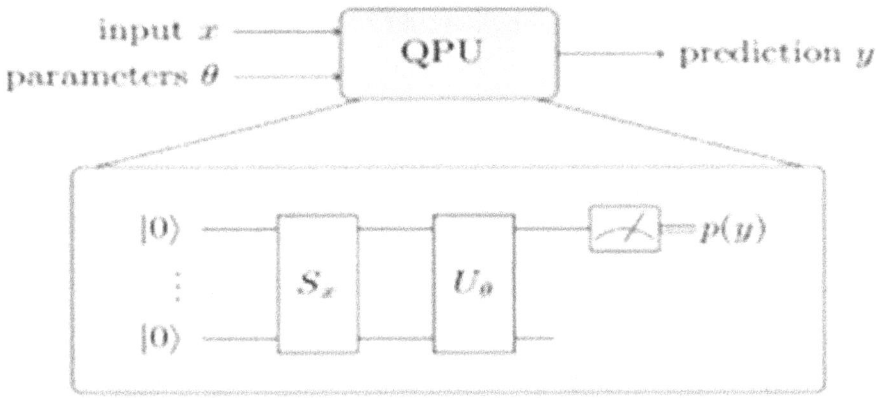

Figure 18 illustrates the potential range of applications for quantum machine learning approaches. To find the lowest energy configurations of molecules for drug development or material science, for example, researchers in the field of quantum chemistry are interested in minimizing high-dimensional and challenging cost functions. These challenges are well suited for quantum computing. It's not unexpected that optimization-based fields like machine learning and quantum chemistry can use the same quantum techniques. As a novel application of quantum computers for ML, the variational quantum Eigen solver employs the same method as the variational classifier. Information learned using machine learning.

Figure 18. General QML architecture

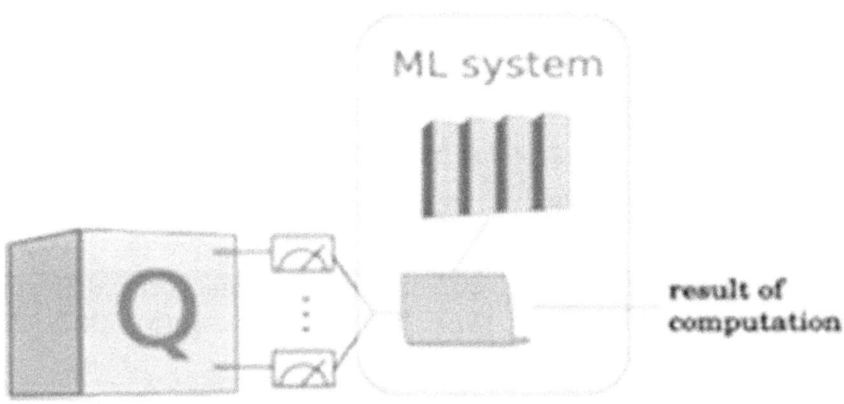

REFERENCES

Aımeur, E., Brassard, G., & Gambs, S. (2006). Machine learning in a quantum world. In *Advances in Artificial Intelligence* (pp. 431–442). Springer.

Arute, F., Arya, K., Babbush, R., Bacon, D., Bardin, J. C., Barends, R., Biswas, R., Boixo, S., & Brandao, F. G. S. L. (2019). Quantum supremacy using a programmable superconducting processor. *Nature*, *574*, 505–510. PMID:31645734

Bausch, J. (2020). Advances in Neural Information Processing Systems: Vol. 33. *Recurrent quantum neural networks.*

Biamonte, J. (2021). Universal variational quantum computation. *Physical Review. A*, *103*(3), L030401.

Biamonte, J., Wittek, P., Pancotti, N., Rebentrost, P., Wiebe, N., & Lloyd, S. (2017). Quantum machine learning. *Nature*, *549*, 195–202. PMID:28905917

Brookes, J. (2017). Quantum effects in biology: Golden rule in enzymes, olfaction, photosynthesis and magneto detection. *Proceedings - Royal Society. Mathematical, Physical and Engineering Sciences*, *473*, 20160822.

Cerezo, M., Arrasmith, A., Babbush, R., Benjamin, S. C., Endo, S., Fujii, K., McClean, J. R., Mitarai, K., Yuan, X., & Cincio, L. (2020). *Variational quantum algorithms*. arXiv preprint arXiv:2012.09265.

Choi, J., Oh, S., & Kim, J. (2020). Energy-efficient cluster head selection via quantum approximate optimization. *Electronics (Basel)*, *9*(10).

Cong, I., Choi, S., & Lukin, M. D. (2019). Quantum convolutional neural networks. *Nature Physics*, *15*(12), 1273–1278.

Deutsch, D. (1985). Quantum theory, the church–turing principle and the universal quantum computer. *Proceedings of the Royal Society of London. Series A, Mathematical and Physical Sciences*, *400*, 97–117.

Dong, D. (2008). Quantum reinforcement learning. *IEEE Transactions on Systems, Man, and Cybernetics. Part B, Cybernetics*, *38*(5), 1207–1220. PMID:18784007

Dong, D., Chen, C., & Li, H. (2020). Quantum reinforcement learning. *Systems, Man, and Cybernetics. IEEE Transactions on, 38*(5), 1207–1220.

Dong, D., Chen, C., Li, H., & Tarn, T.-J. (2008). Quantum reinforcement learning. *IEEE Transactions on Systems, Man, and Cybernetics. Part B, Cybernetics*, *38*(5), 1207–1220. PMID:18784007

Durr, C. & Hoyer, P. (1996). *A Quantum Algorithm for Finding the Minimum*. arXiv: quant - ph / 96070148. https://arxiv.org/abs/quant-ph/9607014.

Farhi, E., Goldstone, J., & Gutmann, S. (2014). *A quantum approximate optimization algorithm*. arXiv preprint arXiv:1411.4028.

Giovannetti, V., Lloyd, S., & Maccone, L. (2008). Quantum Random Access Memory. *Physical Review Letters*, *100*(16), 1484–1509. doi:10.1103/PhysRevLett.100.160501 PMID:18518173

Grover, L. K. (1996). A fast quantum mechanical algorithm for database search. In: *ACM symposium on Theory of computing* (pp. 212– 219). ACM. https://arxiv.org/pdf/quant-ph/9605043. pdf

Jerbi, S., Trenkwalder, L. M., Poulsen Nautrup, H., Briegel, H. J., & Dunjko, V. (2021). Quantum enhancements for deep reinforcement learning in large spaces. *PRX Quantum*, *2*, 010328.

Kandala, A., Mezzacapo, A., Temme, K., Takita, M., Brink, M., Chow, J. M., & Gambetta, J. M. (2017). Hardware-efficient variational quantum eigensolver for small molecules and quantum magnets. *Nature*, *549*(7671), 242–246. PMID:28905916

Kim, J., Kwak, Y., Jung, S., & Kim, J.-H. (2021). Quantum scheduling for millimeter-wave observation satellite constellation. In *Proceedings of the IEEE VTS Asia Pacific Wireless Communications Symposium (APWCS)* (pp. 1–1). IEEE.

Knill, E., LaFlamme, R., & Milburn, G. J. (2001). A scheme for efficient quantum computation with linear optics. *Nature*, *409*, 46–52. PMID:11343107

Koashi, M., Yamamoto, T., & Imoto, N. (2001). Probabilistic manipulation of entangled photons. *Physical Review A*, *63*, 030301.

KopczykD. (2018). *Quantum machine learning for data scientists*. https://arxiv.org/ abs/1804.10068

Li, G., Deng, L., Xu, Y., Wen, C., Wang, Jing Pei, J., & Shi, L. (2016). Temperature based restricted Boltzmann machines. *Scientific Reports, 6*, 19133.

Liu, Y., Jia, Q., & Wang, X. (2022). Quantum reinforcement learning method and application based on value function. *IFAC PapersOnLine,* 132–137.

Lloyd, S., Mohseni, M., & Rebentrost, P. (2013). Quantum algorithms for supervised and unsupervised machine learning. arXiv preprint 1307.0411.

Mishra, P., Sivaram, M., Arvindhan, M., Daniel, A., & Ranjan, R. (2022). Energy-Efficient Cloud Computing Techniques for Next Generation. In Hybrid Intelligent Approaches for Smart Energy (pp. 49–66). Wiley. https://doi.org/10.1002/978111982187 doi: 8.ch3

Mohammad, H. (2018). Amin, Evgeny Andriyash, Jason Rolfe, Bohdan Kulchytskyy, and Roger Melko. Quantum Boltzmann machine. *Physical Review X*, *8*(2), 021050.

Mottonen, M. (2005). Transformation of quantum states using uniformly controlled rotations. *Quantum Information & Computation*, *5*, 467–473.

Niels, M. P. (2023). Quantum reinforcement learning. *Quantum Information Processing*, *22*, 125.

Nielsen, M. & Chuang, I. (2000). Quantum Computation and Quantum Information. Cambridge University Press.

Niemann, DWille, R. (2016). Logic Synthesis for Quantum State Generation. In *IEEE 46th International Symposium on Multiple-Valued Logic*. Springer.

Pittman, T. B., Jacobs, B. C., & Franson, J. D. (2001). Probabilistic quantum logic operations using polarizing Beam splitters. *Physical Review A*, *64*, 062311.

PuskarovT.CuberoA. C. (2020). Machine learning algorithms based on generalized Gibbs ensembles. arXiv:1804.03546

Saharia, A., Maddila, R. K., Ali, J., Yupapin, P., & Singh, G. (2019). An elementary optical logic circuit for quantum computing: A review. *Optical and Quantum Electronics*, *51*, 1–13.

Schuld, M., Fingerhuth, M., & Petruccione, F. (2017). Implementing a distance-based classifier with a quantum interference circuit. *EPL (Europhysics Letters)*. https://arxiv.org/pdf/ 1703.10793.pdf

Schuld, M., & Killoran, N. (2019). Quantum machine learning in feature hilbert spaces. *Physical Review Letters*, *122*(4), 040504. PMID:30768345

Schuld, M., Sinayskiy, I., & Petruccione, F. (2015). An introduction to quantum machine learning. *Contemporary Physics*, *56*, 172–185.

Shor, P. W. (1994). Algorithms for quantum computation: discrete logarithms and factoring. In *Proc. 35th Annual Symp. On Found. Of Computer Science*. IEEE Computer Society.

Wiebe, N., Kapoor, A., & Svore, K. (2014). *Quantum deep learning*. arXiv preprint arXiv:1412.3489.

Wittek, P. (2014). Quantum computing. In Quantum Machine Learning What Quantum Computing Means to Data Mining. Elsevier Inc.

Xia, R. (2018). Sabre Kais Quantum machine learning for electronic structure calculations. *Nature Communications*, *9*, 4195. PMID:30305624

Compilation of References

Abrams, D. S., & Lloyd, S. (1999). Quantum algorithm providing exponential speed increase for finding eigenvalues and eigenvectors. *Physical Review Letters, 83*(24), 5162–5165. doi:10.1103/PhysRevLett.83.5162

Ambainis, A. (2000). Quantum lower bounds by quantum arguments. In *Proceedings of the Thirty-Second Annual ACM Symposium on Theory of Computing,* (pp. 636–643). ACM. 10.1145/335305.335394

Andrew, M. (2003). Childs, Richard Cleve, Enrico Deotto, Edward Farhi, Sam Gutmann, and Daniel A. Spielman. Exponential algorithmic speedup by a quantum walk. In *Proceedings of STOC '03,* (pp. 59–68.). UMD.

Andrew, M. (2007). Childs, Andrew J. Landahl, and Pablo A. Parrilo. Improved quantum algorithms for the ordered search problem via semidefinite programming. *Physical Review A, 75*(3), 032335. doi:10.1103/PhysRevA.75.032335

Andrew, M. (2008). Childs and Troy Lee. Optimal quantum adversary lower bounds for ordered search. *Lecture Notes in Computer Science, 5125*, 869–880. doi:10.1007/978-3-540-70575-8_71

Arunachalam, S., Gheorghiu, V., Jochym-O'Connor, T., Mosca, M., & Srinivasan, P. V. (2015). On the robustness of bucket brigade quantum RAM. *New Journal of Physics, 17*(12), 123010. doi:10.1088/1367-2630/17/12/123010

Arute, F., Arya, K., Babbush, R., Bacon, D., Bardin, J. C., Barends, R., Biswas, R., Boixo, S., & Brandao, F. G. S. L. (2019). Quantum supremacy using a programmable superconducting processor. *Nature, 574*, 505–510. PMID:31645734

Asshoff, P., Merz, A., Kalt, H., & Hetterich, M. (2011). A spintronic source of circularly polarized single photons. *Applied Physics Letters, 98*(11), 112106. doi:10.1063/1.3564893

Atkins, P. W. & Friedman, R. S. (2010). *Molecular Quantum Mechanics.* Oxford university press, UK.

Atkins, P. W. & Friedman, R. S. (2010). *Molecular Quantum Mechanics.* Oxford University Press.

Bausch, J. (2020). Advances in Neural Information Processing Systems: Vol. 33. *Recurrent quantum neural networks.*

Bergholm, V., Izaac, J., Schuld, M., Gogolin, C., Blank, C., McKiernan, K., & Killoran, N. (2018). *Pennylane: Automatic differentiation of hybrid quantum-classical computations.* arXiv preprint arXiv:1811.04968

Biamonte, J. (2021). Universal variational quantum computation. *Physical Review. A, 103*(3), L030401.

Biamonte, J., Wittek, P., Pancotti, N., Rebentrost, P., Wiebe, N., & Lloyd, S. (2017). Quantum machine learning. *Nature, 549*, 195 202. PMID:28905917

Biham, E., Biham, O., Biron, D., Grassl, M., & Lidar, D. A. (1999). Grover's quantum search algorithm for an arbitrary initial amplitude distribution. *Physical Review A, 60*(4), 2742–2745. doi:10.1103/PhysRevA.60.2742

Bonner, R., & Freivalds, R. (2003). A survey of quantum learning. Quantum Computation and Learning, 106.

Brassard, G., Høyer, P., Mosca, M., & Tapp, A. (2000). Quantum amplitude amplification and estimation. *Contemporary Mathematics, 305*, 53–74. doi:10.1090/conm/305/05215

Brian L.F. (2004). *Matlab tutor CD.* John Wiley & Sons, (UK).

Brookes, J. (2017). Quantum effects in biology: Golden rule in enzymes, olfaction, photosynthesis and magneto detection. *Proceedings - Royal Society. Mathematical, Physical and Engineering Sciences, 473*, 20160822.

Broughton, M., Verdon, G., McCourt, T., Martinez, A. J., Yoo, J. H., Isakov, S. V., Massey, P., & Niu, Y. (2020). *Tensorflow quantum: a software framework for quantum machine learning.* arXiv preprint arXiv:2003.02989.

Bryant G.W. & Solomon G.S. (2004). *Optics of Quantum dots and wires.* Artech house, Inc. MA, US.

Bryant, G.W, & Solomon, G.S. (2004). *Optics of Quantum dots and wires.* Artech House, Inc.

Cai, Z., Babbush, R., Benjamin, S., Huggins, W., Li, Y., McClean, J., & O'Brien. (2022). *Quantum Error mitigation.* Cornell University.

Carleo, G., & Troyer, M. (2017). Solving the quantum many-body problem with artificial neural networks. *Science, 355*(6325), 602–606. doi:10.1126cience.aag2302 PMID:28183973

Carrasquilla, J., & Melko, R. G. (2017). Machine learning phases of matter. *Nature Physics, 13*(5), 431–434. doi:10.1038/nphys4035

Cerezo, M., Arrasmith, A., Babbush, R., Benjamin, S. C., Endo, S., Fujii, K., McClean, J. R., Mitarai, K., Yuan, X., & Cincio, L. (2020). *Variational quantum algorithms*. arXiv preprint arXiv:2012.09265.

Childs, A., Farhi, E., Goldstone, J., & Gutmann, S. (2002). Finding cliques by quantum adiabatic evolution. *Quantum Information & Computation*, *2*(181), 181–191. doi:10.26421/QIC2.3-1

Childs, A., Farhi, E., & Preskill, J. (2001). Robustness of adiabatic quantum computation. *Physical Review A*, *65*(1), 012322. doi:10.1103/PhysRevA.65.012322

Choi, J., Oh, S., & Kim, J. (2020). Energy-efficient cluster head selection via quantum approximate optimization. *Electronics (Basel)*, *9*(10).

Cleve B. Moler, C. (2004). *Numerical Computing with Matlab.* Siam Publishing.

Cleve B. Moler. (2004). *Numerical Computing with Matlab.* Siam Publishing, Philadelphia.

CleveR. (1999). *An introduction to quantum complexity theory*. arXiv:quant-ph/9906111v1.

Cleve, R., Gottesman, D., & Lo, H.-K. (1999). How to share a quantum secret. *Physical Review Letters*, *83*(3), 648–651. doi:10.1103/PhysRevLett.83.648

Coccia, M., Roshani, S., & Mosleh, M. (2022). Evolution of quantum computing: Theoretical and innovation management implications for emerging quantum industry. *IEEE Transactions on Engineering Management*, 1–11. doi:10.1109/TEM.2022.3175633

Cong, I., Choi, S., & Lukin, M. D. (2019). Quantum convolutional neural networks. *Nature Physics*, *15*(12), 1273–1278.

Coppersmith, D. (1994). *An approximate Fourier transform is useful in quantum factoring. Research Report RC 19642*. IBM.

Crépeau, C., Gottesman, D., & Smith, A. (2002). Secure multi-party quantum computation. *Proceedings of STOC '02*, 643–652. 10.1145/509907.510000

Cross, A., DiVincenzo, D. P., & Terhal, B. (2007). A comparative code study for quantum fault tolerance. arXiv:quant-ph/0711.1556v1.

Daku, B. (2006). *Matlab tutor CD.* John Wiley & Sons.

Datta, S., & Das, B. (1989). Electronic analog of the electrooptic modulator. *Applied Physics Letters*, *56*(7), 665–667. doi:10.1063/1.102730

David Clader, B., Jacobs, B. C., & Sprouse, C. R. (2013). Preconditioned quantum linear system algorithm. *Physical Review Letters*, *110*(25), 250504. doi:10.1103/PhysRevLett.110.250504 PMID:23829722

Deng, D.-L., Li, X., & Das Sarma, S. (2017). Quantum entanglement in neural network states. *Physical Review X*, *7*(2), 021021. doi:10.1103/PhysRevX.7.021021

Deutsch, D. (1985). Quantum theory, the church–turing principle and the universal quantum computer. *Proceedings of the Royal Society of London. Series A, Mathematical and Physical Sciences*, *400*, 97–117.

Domingos, P. (2012). Afewuseful things to knowabout machine learning. *Communications of the ACM*, *55*(10), 78–87. doi:10.1145/2347736.2347755

Dong, D., Chen, C., & Li, H. (2020). Quantum reinforcement learning. *Systems, Man, and Cybernetics. IEEE Transactions on*, *38*(5), 1207–1220.

Dong, D. (2008). Quantum reinforcement learning. *IEEE Transactions on Systems, Man, and Cybernetics. Part B, Cybernetics*, *38*(5), 1207–1220. PMID:18784007

Dong, D., Chen, C., Li, H., & Tarn, T.-J. (2008). Quantum reinforcement learning. *IEEE Transactions on Systems, Man, and Cybernetics. Part B, Cybernetics*, *38*(5), 1207–1220. PMID:18784007

Durr, C. & Hoyer, P. (1996). *A Quantum Algorithm for Finding the Minimum*. arXiv: quant - ph / 96070148. https://arxiv.org/abs/quant-ph/9607014.

Dwivedi, A., Saini, G. K., & Musa, U. I. (2023, March). Cybersecurity and Prevention in the Quantum Era. In *2023 2nd International Conference for Innovation in Technology (INOCON)* (pp. 1-6). IEEE. 10.1109/INOCON57975.2023.10101186

Farhi, E., Goldstone, J., & Gutmann, S. (2014). *A quantum approximate optimization algorithm*. arXiv preprint arXiv:1411.4028.

Farhi, E., Goldstone, J., & Gutmann, S. (2014). *A quantum approximate optimization algorithm*. https://arxiv.org/abs/1411.4028

Farhi, E., Goldstone, J., Gutmann, S., & Sipser, M. (2000). Quantum computation by adiabatic evolution. arXiv:quant-ph/0001106. MIT-CTP-2936.

FarhiE.GoldstoneJ.GutmannS. (2014). A quantum approximate optimization algorithm. arXiv:1411.4028

FarhiE.HarrowA. W. (2016). Quantum supremacy through the quantum approximate optimization algorithm. arXiv:1602.07674

Fernández-Caramés, T. M. (2019). From pre-quantum to post-quantum IoT security: A survey on quantum-resistant cryptosystems for the Internet of Things. *IEEE Internet of Things Journal*, *7*(7), 6457–6480. doi:10.1109/JIOT.2019.2958788

Feynman, R. P., Hibbs, A. R., & Styer, D. F. (2010). *Quantum mechanics and path integrals*. Courier Corporation.

Fox, M. (2010). *Optical properties of Solids.* Oxford University Press.

Gao, X., Wang, Y., Chen, X., & Gao, S. (2021). Interface, interaction, and intelligence in generalized brain–computer interfaces. *Trends in Cognitive Sciences*, *25*(8), 671–684. doi:10.1016/j.tics.2021.04.003 PMID:34116918

Giacomo, T., Guglielmo, M., Juan, C., Matthias, T., Roger, M., & Giuseppe, C. (2018). Neural-network quantum state tomography. *Nature Physics*, *14*(5), 447–450. doi:10.103841567-018-0048-5

Gilles Brassard, E. A., & Gambs, S. (2006). Machine learning in a quantum world. In *Advances in Artificial Intelligence* (pp. 431–442). Springer.

Giovannetti, V., Lloyd, S., & Maccone, L. (2008). Quantum random access memory. *Physical Review Letters*, *100*(16), 160501. doi:10.1103/PhysRevLett.100.160501 PMID:18518173

Glasser, I., Pancotti, N., & Cirac, J. N. (2018). Supervised learning with generalized tensor networks. arXiv preprint arXiv:1806.05964.

Graham, P. (2006, April). Collins. Computing with quantum knots. *Scientific American*, *294*(4), 56–63. doi:10.1038cientificamerican0406-56 PMID:16596880

Grover, L. K. (1996). A fast quantum mechanical algorithm for database search. In: *ACM symposium on Theory of computing* (pp. 212– 219). ACM. https://arxiv.org/pdf/quant-ph/9605043. pdf

Hallgren, S., Russell, A., & Ta-Shma, A. (2000). Normal subgroup reconstruction and quantum computing using group representations. In *Proceedings of STOC '00*, (pp. 627–635). ACM. 10.1145/335305.335392

Hallgren, S., Russell, A., & Ta-Shma, A. (2003). The hidden subgroup problem and quantum computation using group representations. *SIAM Journal on Computing*, *32*(4), 916–934. doi:10.1137/S009753970139450X

Hanselman, D & Littlefield, B. (2000). *Mastering Matlab 6 – A Comprehensive tutorial and reference.* Prentice hall, NJ.

Hanselman, D & Littlefield, B. (2000). *Mastering Matlab 6 – A Comprehensive tutorial and reference.* Prentice Hall.

Hardy, G. H., & Wright, E. M. (1979). *An Introduction to the Theory of Numbers*. Oxford University Press.

Harrison, P. (1988). Quantum Wells, Wires and Dots: Theoretical and Computational Physics. John Wiley & Sons, (UK).

Harrison, P. (1997). *Computational methods in Physics, Chemistry & Biology: An Introduction.* John Wiley & Sons, (UK).

Harrison, P. (2001). *Computational methods in Physics, Chemistry & Biology: An Introduction.* John Wiley & Sons.

Harrison, P. (2016). *Quantum Wells, Wires and Dots: Theoretical and Computational Physics.* John Wiley & Sons.

Hettrich, M. (2010). *Electrical spin injection into single InGaAs quantum dots*. Institute of Applied Physics, KIT.

Hillery, M., Buzek, V., & Berthiaume, A. (1999). Quantum secret sharing. *Physical Review A*, *59*(3), 1829–1834. doi:10.1103/PhysRevA.59.1829

Hinchliffe, A. (1988). *Computational Quantum Chemistry*. John Wiley & Sons.

Huang, H., Xu, X., Guo, C., Tian, G., Wei, S., Sun, X., Bao, W., & Long, G. (2021). *Quantum Physics.* Cornell University.

Ignacio, C. & Zoller, P. (1995). Quantum computations with cold trapped ions. *Physical Review Letters, 74,* 4091-4094. doi:10.1103/PhysRevLett.74.4091 PMID:10058410

Isaac, L. (1997). Chuang and Michael Nielsen. Prescription for experimental determination of the dynamics of a quantum black box. *Journal of Modern Optics*, *44*, 2567–2573.

Islam, M., Chowdhury, M., Khan, Z., & Khan, S. M. (2022). Hybrid quantum-classical neural network for cloud-supported in-vehicle cyberattack detection. *IEEE Sensors Letters*, *6*(4), 1–4. doi:10.1109/LSENS.2022.3153931

Ivo Timon, V. I. N. K. (2008). *Manipulation and Read-out of Spins in Quantum Dots*. Technische Universiteit Delft.

Jacak, L. & Wojs, H. (1998). *Quantum dots*. Springer, Berlin.

Jacak, L. & Wojs. H. (1998). *Quantum dots*. Springer.

Jadhav, A., Rasool, A., & Gyanchandani, M. (2023). Quantum Machine Learning: Scope for real-world problems. *Procedia Computer Science*, *218*, 2612–2625. doi:10.1016/j.procs.2023.01.235

James, W. (1965). Cooley and John W. Tukey. An algorithm for the machine calculation of complex Fourier series. *Mathematics of Computation*, *19*(90), 297–301. doi:10.1090/S0025-5718-1965-0178586-1

Jerbi, S., Trenkwalder, L. M., Poulsen Nautrup, H., Briegel, H. J., & Dunjko, V. (2021). Quantum enhancements for deep reinforcement learning in large spaces. *PRX Quantum*, *2*, 010328.

Kalinin, M., & Krundyshev, V. (2023). Security intrusion detection using quantum machine learning techniques. *Journal of Computer Virology and Hacking Techniques*, *19*(1), 125–136. doi:10.100711416-022-00435-0

Kamaleswari, P., & Krishnaraj, N. (2023, February). An Assessment of Object Detection in Thermal (Infrared) Image Processing. In *2023 Third International Conference on Artificial Intelligence and Smart Energy (ICAIS)* (pp. 1498-1503). IEEE.

Kandala, A., Mezzacapo, A., Temme, K., Takita, M., Brink, M., Chow, J. M., & Gambetta, J. M. (2017). Hardware-efficient variational quantum eigensolver for small molecules and quantum magnets. *Nature*, *549*(7671), 242–246. PMID:28905916

Khan, T. M., & Robles-Kelly, A. (2020). Machine learning: Quantum vs classical. *IEEE Access : Practical Innovations, Open Solutions*, *8*, 219275–219294. doi:10.1109/ACCESS.2020.3041719

Kim, J., Oh, B., Chong, Y., Hwang, E., & Park, D. (2021). *Quantum readout error mmitigation via deep learning.* Cornell University.

Kim, J., Kwak, Y., Jung, S., & Kim, J.-H. (2021). Quantum scheduling for millimeter-wave observation satellite constellation. In *Proceedings of the IEEE VTS Asia Pacific Wireless Communications Symposium (APWCS)* (pp. 1–1). IEEE.

Kimmel, S., Lin, C. Y.-Y., Low, G. H., Ozols, M., & Yoder, T. J. (2016). *Hamiltonian simulation with optimal sample complexity*. https://arxiv.org/abs/1608.00281

Knight, W. (2005, February). 'Quantum well' transistor promises lean computing. *New Scientist*, 10.

Knill, E., LaFlamme, R., & Milburn, G. J. (2001). A scheme for efficient quantum computation with linear optics. *Nature*, *409*, 46–52. PMID:11343107

Koashi, M., Yamamoto, T., & Imoto, N. (2001). Probabilistic manipulation of entangled photons. *Physical Review A*, *63*, 030301.

Kontsevich, M., & Manin, Y. (1994). Gromov-Witten classes, quantum cohomology, and enumerative geometry. *Communications in Mathematical Physics*, *164*(3), 525–562. doi:10.1007/BF02101490

KopczykD. (2018). *Quantum machine learning for data scientists*. https://arxiv.org/abs/1804.10068

Le, Q. V. (2013). Building high-level features using large scale unsupervised learning. In *Acoustics, Speech and Signal Processing (ICASSP), IEEE International Conference on,* (pp. 8595–8598). IEEE. 10.1109/ICASSP.2013.6639343

Leskowitz, G. M., & Mueller, L. J. (2004). State interrogation in nuclear magnetic resonance quantum-information processing. *Physical Review A*, *69*(5), 052302. doi:10.1103/PhysRevA.69.052302

Levi, A. (2012). *Applied Quantum Mechanics* (2nd ed.). Cambridge University Press.

Levi, A. (2012). *Applied Quantum Mechanics (2nd edn.)*. Cambridge university press, UK.

Li, G., Deng, L., Xu, Y., Wen, C., Wang, Jing Pei, J., & Shi, L. (2016). Temperature based restricted Boltzmann machines. *Scientific Reports, 6*, 19133.

Liboff, R. (2002). *Introductory Quantum mechanics (4th edn.)*. Pearson education Inc.

Liboff, R. (2008). *Introductory Quantum mechanics (4th edn.)*. Pearson education Inc.

Lin, J., Wallman, J. J., Hincks, I., & Laflamme, R. (2021). Independent state and measurement characterization for quantum computers. In Physical Review Research, 3(3). American Physical Society (APS). doi:10.1103/PhysRevResearch.3.033285

Liu, Y., Jia, Q., & Wang, X. (2022). Quantum reinforcement learning method and application based on value function. *IFAC PapersOnLine,* 132–137.

Li, Y., Tian, M., Liu, G., Peng, C., & Jiao, L. (2020). Quantum optimization and quantum learning: A survey. *IEEE Access : Practical Innovations, Open Solutions, 8*, 23568–23593. doi:10.1109/ACCESS.2020.2970105

Lloyd, S. (1996). Universal quantum simulators. Science 273, 1073–1078. doi:10.1126cience.273.5278.1073

Lloyd, S., Mohseni, M., & Rebentrost, P. (2013). Quantum algorithms for supervised and unsupervised machine learning. arXiv preprint 1307.0411.

Lloyd, S., Mohseni, M., & Rebentrost, P. (2013). Quantum algorithms for supervised and unsupervised machine learning. arXiv preprint arXiv:1307.0411

Loeffler, W. (2008). *Electrical preparation of spin-polarized electrons in semiconductor quantum dots*. Institute of Applied Physics, KIT.

Löffler, W., Höpcke, N., Kalt, H., Li, S. F., Grün, M., & Hetterich, M. (2010). Doping and optimal electron spin polarization in n-ZnMnSe for quantum-dot spin-injection light-emitting diodes. *Applied Physics Letters*, *96*(5), 052113. doi:10.1063/1.3308500

Luo, X.-Z., Liu, J.-G., Zhang, P., Wang, L., & Yao, J. (2019). Extensible, efficient framework for quantum algorithm design. arXiv preprint arXiv:1912.10877.

Merz, A. (2014). *Fast electron spin resonance controlled manipulation of spin injection into quantum dots*. KIT. doi:10.1063/1.4884016

Michler, P. (2003). *Single quantum dots: Fundamentals, applications and new concepts.* Springer, Berlin.

Michler, P. (2003). *Single quantum dots: Fundamentals, applications and new concepts.* Springer.

Miguel Arrazola, J., Delgado, A., Bardhan, B. R., & Lloyd, S. (2019). *Quantum-inspired algorithms in practice.* arXiv preprint arXiv:1905.10415.

Mishra, P., Sivaram, M., Arvindhan, M., Daniel, A., & Ranjan, R. (2022). Energy-Efficient Cloud Computing Techniques for Next Generation. In Hybrid Intelligent Approaches for Smart Energy (pp. 49–66). Wiley. https://doi.org/10.1002/978111982187 doi: 8.ch3

Mohammad, H. (2018). Amin, Evgeny Andriyash, Jason Rolfe, Bohdan Kulchytskyy, and Roger Melko. Quantum Boltzmann machine. *Physical Review X*, *8*(2), 021050.

Mottonen, M. (2005). Transformation of quantum states using uniformly controlled rotations. *Quantum Information & Computation*, *5*, 467–473.

Neven, H., Denchev, V. S., Rose, G., & Macready, W. G. (2009). Training a large scale classifier with the quantum adiabatic algorithm. arXiv preprint arXiv:0912.0779.

Nielsen, M. & Chuang, I. (2000). Quantum Computation and Quantum Information. Cambridge University Press.

Nielsen, M. A., & Chuang, I. L. (2010). *Quantum Computation and Quantum Information.* Cambridge University Press.

Niels, M. P. (2023). Quantum reinforcement learning. *Quantum Information Processing*, *22*, 125.

Niemann, DWille, R. (2016). Logic Synthesis for Quantum State Generation. In *IEEE 46th International Symposium on Multiple-Valued Logic.* Springer.

Peruzzo, A., McClean, J., Shadbolt, P., Yung, M.-H., Zhou, X.-Q., Love, P. J., Aspuru-Guzik, A., & Obrien, J. L. (2014). A variational eigenvalue solver on a photonic quantum processor. *Nature Communications*, *5*(1), 5. doi:10.1038/ncomms5213 PMID:25055053

Pittman, T. B., Jacobs, B. C., & Franson, J. D. (2001). Probabilistic quantum logic operations using polarizing Beam splitters. *Physical Review A*, *64*, 062311.

Pratap, R. (2009). *Getting started with Matlab – A quick introduction for Scientists and Engineers.* Oxford university press, UK.

Pratap, R. (2009). *Getting started with Matlab – A quick introduction for Scientists and Engineers.* Oxford University Press.

PuskarovT.CuberoA. C. (2020). Machine learning algorithms based on generalized Gibbs ensembles. arXiv:1804.03546

Quintero-Torres, R., Vázquez-Cerón, E., Rodríguez-Rodríguez, E., Stintz, A. and Diels, J.-C. (2005). Multiple quantum wells for passive ultra short laser pulse generation. phys. stat. sol. (c), 2: 3015-3018.

Ralegankar, V. K., Bagul, J., Thakkar, B., Gupta, R., Tanwar, S., Sharma, G., & Davidson, I. E. (2021). Quantum cryptography-as-a-service for secure UAV communication: Applications, challenges, and case study. *IEEE Access : Practical Innovations, Open Solutions*, *10*, 1475–1492. doi:10.1109/ACCESS.2021.3138753

Ramachandra, A., & Singh, M. (2022). *A Hybrid Quantum-Classical Neural Network for Image Classification*. Innova Solutions.

Ranjan, A., Arya, A. K., & Ravinder, M. (2020, December). Quantum Techniques for Image Processing. In *2020 2nd International Conference on Advances in Computing, Communication Control and Networking (ICACCCN)* (pp. 1035-1039). IEEE. 10.1109/ICACCCN51052.2020.9362910

Rasool, R. U., Ahmad, H. F., Rafique, W., Qayyum, A., & Qadir, J. (2022). Quantum computing for healthcare. *RE:view*.

Rebentrost, P., Mohseni, M., & Lloyd, S. (2014). Quantum support vector machine for big data classification. *Physical Review Letters*, *113*(13), 130503. doi:10.1103/PhysRevLett.113.130503 PMID:25302877

RebentrostP.SteffensA.LloydS. (2016). *Quantum singular value decomposition of non-sparse low-rank matrices*. https://arxiv.org/abs/1607.05404

Ren, Z. A., Chen, Y. P., Liu, J. Y., Ding, H. J., & Wang, Q. (2020). Implementation of machine learning in quantum key distributions. *IEEE Communications Letters*, *25*(3), 940–944. doi:10.1109/LCOMM.2020.3040212

Rosenblatt, F. (1958). The perceptron: A probabilistic model for information storage and organization in the brain. *Psychological Review*, *65*(6), 386–408. doi:10.1037/h0042519 PMID:13602029

Sachdeva, K., Sachdeva, R., & Gupta, H. (2023). Quantum Computing in Image Processing. In Recent Developments in Electronics and Communication Systems (pp. 25-30). IOS Press. doi:10.3233/ATDE221232

Saharia, A., Maddila, R. K., Ali, J., Yupapin, P., & Singh, G. (2019). An elementary optical logic circuit for quantum computing: A review. *Optical and Quantum Electronics*, *51*, 1–13.

Scherer, A. (2015). *Resource analysis of the quantum linear system algorithm*. https://arxiv.org/abs/1505.06552

Schuld, M., Fingerhuth, M., & Petruccione, F. (2017). Implementing a distance-based classifier with a quantum interference circuit. *EPL (Europhysics Letters)*. https://arxiv.org/pdf/ 1703.10793.pdf

Schuld, M., & Killoran, N. (2019). Quantum machine learning in feature hilbert spaces. *Physical Review Letters*, *122*(4), 040504. PMID:30768345

Schuld, M., Sinayskiy, I., & Petruccione, F. (2014). The quest for a quantum neural network. *Quantum Information Processing*, *13*(11), 2567–2586. doi:10.100711128-014-0809-8

Schuld, M., Sinayskiy, I., & Petruccione, F. (2015). An introduction to quantum machine learning. *Contemporary Physics*, *56*, 172–185.

Schuld, M., Sinayskiy, I., & Petruccione, F. (2015). Introduction to quantum machine learning. *Contemporary Physics*, *56*(2), 172–185. doi:10.1080/00107514.2014.964942

Schuld, M., Sinayskiy, I., & Petruccione, F. (2016). Prediction by linear regression on a quantum computer. *Physical Review. A*, *94*(2), 022342. doi:10.1103/PhysRevA.94.022342

Shor, P. W. (1994). Algorithms for quantum computation: discrete logarithms and factoring. In *Proc. 35th Annual Symp. On Found. Of Computer Science.* IEEE Computer Society.

Simões, R. D. M., Huber, P., Meier, N., Smailov, N., Füchslin, R. M., & Stockinger, K. (2023). Experimental Evaluation of Quantum Machine Learning Algorithms. *IEEE Access : Practical Innovations, Open Solutions*, *11*, 6197–6208. doi:10.1109/ACCESS.2023.3236409

Singh, A., Dev, K., Siljak, H., Joshi, H. D., & Magarini, M. (2021). Quantum internet—Applications, functionalities, enabling technologies, challenges, and research directions. *IEEE Communications Surveys and Tutorials*, *23*(4), 2218–2247. doi:10.1109/COMST.2021.3109944

Singh, J., & Bhangu, K. S. (2023). Contemporary Quantum Computing Use Cases: Taxonomy, Review and Challenges. *Archives of Computational Methods in Engineering*, *30*(1), 615–638. doi:10.100711831-022-09809-5

Srikanth, P., & Kumar, A. (2022). *Secure Quantum Computing for Healthcare Sector: A Short Analysis*. arXiv preprint arXiv:2211.10027.

Stoudenmire, E., & Schwab, D. J. (2016). Supervised learning with tensor networks. Advances in Neural Information Processing Systems, 4799–4807.

Tacchino, F., Mangini, S., Barkoutsos, P. K., Macchiavello, C., Gerace, D., Tavernelli, I., & Bajoni, D. (2021). Variational learning for quantum artificial neural networks. *IEEE Transactions on Quantum Engineering*, *2*, 1–10. doi:10.1109/TQE.2021.3062494

Tang, E. (2019). Aquantum-inspired classical algorithm for recommendation systems. In *Proceedings of the 51st Annual ACM SIGACT Symposium on Theory of Computing* (pp. 217–228). ACM. 10.1145/3313276.3316310

Thijssen, J. (2007). *Computational Physics (2nd edn.).* Cambridge university press, UK.

Thijssen. J. M. (2007). *Computational Physics* (2nd ed.). Cambridge University Press.

Thomas, H. (2001). *Introduction to Algorithms*. MIT Press.

Ullah, U., Jurado, A. G. O., Gonzalez, I. D., & Garcia-Zapirain, B. (2022). A Fully Connected Quantum Convolutional Neural Network for Classifying Ischemic Cardiopathy. *IEEE Access : Practical Innovations, Open Solutions, 10*, 134592–134605. doi:10.1109/ACCESS.2022.3232307

Van Loan, C. (1996) *Introduction to scientific computation – A Matrix vector approach using Matlab.* Prentice Hall.

Van Loan, C. (1996). *Introduction to scientific computation – A Matrix vector approach using Matlab.* Prentice hall, NJ

Ventura, D., & Martinez, T. (2000). Quantum associative memory. *Information Sciences, 124*(1), 273–296. doi:10.1016/S0020-0255(99)00101-2

VerdonG.BroughtonM.BiamonteJ. (2017). A quantum algorithm to train neural networks using low-depth circuits. arXiv:1712.05304

Vitalii, Y. (2010). ODMR study of Zn1-xMnxSe/Zn1-yBeySe and (Cd1-x,Mn)Te/Cd1-yMgyTe Diluted Magnetic semiconductor quantum wells. *Applied Magnetic Resonance, 39*(1-2), 31–47. doi:10.100700723-010-0133-0

Vukmirovic, N., Gacevic, Z., Ikonic, Z., Injin, D., Harrison, P., & Milanovic, V. (2006, July). Intraband absorption in InAs/GaAs quantum dot infrared photodetectors –effective mass versus k x p modelling. *Semiconductor Science and Technology*, 3.

Watrous, J. (2018). *Theory of Quantum Information.* Cambridge University Press. doi:10.1017/9781316848142

Weiss, P. (2006). Quantum-Dot Leap. *Science News Week, 169*(22), 344.

Weiss, P. (2006). Quantum-Dot Leap. *Science News Week, 169*(22).

Whitfield, J. D., Faccin, M., & Biamonte, J. D. (2012). Ground-state spin logic. *Europhysics Letters, 99*, 57004. doi:10.1209/0295-5075/99/57004

Wiebe, N., Kapoor, A., & Svore, K. (2014). *Quantum deep learning.* arXiv preprint arXiv:1412.3489.

Wittek, P. (2014). Quantum computing. In Quantum Machine Learning What Quantum Computing Means to Data Mining. Elsevier Inc.

Wittek, P. (2014). *Quantum Machine Learning: What Quantum Computing Means to Data Mining.* Academic Press.

Wolpaw, J. R. (2007, October). Brain-computer interfaces (BCIs) for communication and control. In *Proceedings of the 9th international ACM SIGACCESS conference on Computers and Accessibility* (pp. 1-2). 10.1145/1296843.1296845

Wong, S. (1997). *Computational methods in Physics & Engineering* (2nd ed.). World Scientific.

Xia, R. (2018). Sabre Kais Quantum machine learning for electronic structure calculations. *Nature Communications*, *9*, 4195. PMID:30305624

Xie, Z. G., Götzinger, S., Fang, W., Cao, H., & Solomon, G. S. (2007). Influence of a single quantum dot state on the characteristics of a microdisk laser. *Physical Review Letters*, *98*(11), 117401. doi:10.1103/PhysRevLett.98.117401 PMID:17501091

Yang, Z., Zolanvari, M., & Jain, R. (2023). A Survey of Important Issues in Quantum Computing and Communications. *IEEE Communications Surveys and Tutorials*, *25*(2), 1059–1094. doi:10.1109/COMST.2023.3254481

Zeng, Y., Xu, P., He, X., Liu, Y., Liu, M., Wang, J., Papoular, D. J., Shlyapnikov, G. V., & Zhan, M. (2017). Entangling two individual atoms of different isotopes via Rydberg blockade. *Physical Review Letters*, *119*(16), 160502. doi:10.1103/PhysRevLett.119.160502 PMID:29099205

Related References

To continue our tradition of advancing academic research, we have compiled a list of recommended IGI Global readings. These references will providc additional information and guidance to further enrich your knowledge and assist you with your own research and future publications.

Abbasnejad, B., Moeinzadeh, S., Ahankoob, A., & Wong, P. S. (2021). The Role of Collaboration in the Implementation of BIM-Enabled Projects. In J. Underwood & M. Shelbourn (Eds.), *Handbook of Research on Driving Transformational Change in the Digital Built Environment* (pp. 27–62). IGI Global. https://doi.org/10.4018/978-1-7998-6600-8.ch002

Abdulrahman, K. O., Mahamood, R. M., & Akinlabi, E. T. (2022). Additive Manufacturing (AM): Processing Technique for Lightweight Alloys and Composite Material. In K. Kumar, B. Babu, & J. Davim (Ed.), *Handbook of Research on Advancements in the Processing, Characterization, and Application of Lightweight Materials* (pp. 27-48). IGI Global. https://doi.org/10.4018/978-1-7998-7864-3.ch002

Agrawal, R., Sharma, P., & Saxena, A. (2021). A Diamond Cut Leather Substrate Antenna for BAN (Body Area Network) Application. In V. Singh, V. Dubey, A. Saxena, R. Tiwari, & H. Sharma (Eds.), *Emerging Materials and Advanced Designs for Wearable Antennas* (pp. 54–59). IGI Global. https://doi.org/10.4018/978-1-7998-7611-3.ch004

Ahmad, F., Al-Ammar, E. A., & Alsaidan, I. (2022). Battery Swapping Station: A Potential Solution to Address the Limitations of EV Charging Infrastructure. In M. Alam, R. Pillai, & N. Murugesan (Eds.), *Developing Charging Infrastructure and Technologies for Electric Vehicles* (pp. 195–207). IGI Global. doi:10.4018/978-1-7998-6858-3.ch010

Aikhuele, D. (2018). A Study of Product Development Engineering and Design Reliability Concerns. *International Journal of Applied Industrial Engineering*, *5*(1), 79–89. doi:10.4018/IJAIE.2018010105

Al-Khatri, H., & Al-Atrash, F. (2021). Occupants' Habits and Natural Ventilation in a Hot Arid Climate. In R. González-Lezcano (Ed.), *Advancements in Sustainable Architecture and Energy Efficiency* (pp. 146–168). IGI Global. https://doi.org/10.4018/978-1-7998-7023-4.ch007

Al-Shebeeb, O. A., Rangaswamy, S., Gopalakrishan, B., & Devaru, D. G. (2017). Evaluation and Indexing of Process Plans Based on Electrical Demand and Energy Consumption. *International Journal of Manufacturing, Materials, and Mechanical Engineering*, *7*(3), 1–19. doi:10.4018/IJMMME.2017070101

Amuda, M. O., Lawal, T. F., & Akinlabi, E. T. (2017). Research Progress on Rheological Behavior of AA7075 Aluminum Alloy During Hot Deformation. *International Journal of Materials Forming and Machining Processes*, *4*(1), 53–96. doi:10.4018/IJMFMP.2017010104

Amuda, M. O., Lawal, T. F., & Mridha, S. (2021). Microstructure and Mechanical Properties of Silicon Carbide-Treated Ferritic Stainless Steel Welds. In L. Burstein (Ed.), *Handbook of Research on Advancements in Manufacturing, Materials, and Mechanical Engineering* (pp. 395–411). IGI Global. https://doi.org/10.4018/978-1-7998-4939-1.ch019

Anikeev, V., Gasem, K. A., & Fan, M. (2021). Application of Supercritical Technologies in Clean Energy Production: A Review. In L. Chen (Ed.), *Handbook of Research on Advancements in Supercritical Fluids Applications for Sustainable Energy Systems* (pp. 792–821). IGI Global. https://doi.org/10.4018/978-1-7998-5796-9.ch022

Arafat, M. Y., Saleem, I., & Devi, T. P. (2022). Drivers of EV Charging Infrastructure Entrepreneurship in India. In M. Alam, R. Pillai, & N. Murugesan (Eds.), *Developing Charging Infrastructure and Technologies for Electric Vehicles* (pp. 208–219). IGI Global. https://doi.org/10.4018/978-1-7998-6858-3.ch011

Araujo, A., & Manninen, H. (2022). Contribution of Project-Based Learning on Social Skills Development: An Industrial Engineer Perspective. In A. Alves & N. van Hattum-Janssen (Eds.), *Training Engineering Students for Modern Technological Advancement* (pp. 119–145). IGI Global. https://doi.org/10.4018/978-1-7998-8816-1.ch006

Armutlu, H. (2018). Intelligent Biomedical Engineering Operations by Cloud Computing Technologies. In U. Kose, G. Guraksin, & O. Deperlioglu (Eds.), *Nature-Inspired Intelligent Techniques for Solving Biomedical Engineering Problems* (pp. 297–317). Hershey, PA: IGI Global. doi:10.4018/978-1-5225-4769-3.ch015

Atik, M., Sadek, M., & Shahrour, I. (2017). Single-Run Adaptive Pushover Procedure for Shear Wall Structures. In V. Plevris, G. Kremmyda, & Y. Fahjan (Eds.), *Performance-Based Seismic Design of Concrete Structures and Infrastructures* (pp. 59–83). Hershey, PA: IGI Global. doi:10.4018/978-1-5225-2089-4.ch003

Attia, H. (2021). Smart Power Microgrid Impact on Sustainable Building. In R. González-Lezcano (Ed.), *Advancements in Sustainable Architecture and Energy Efficiency* (pp. 169–194). IGI Global. https://doi.org/10.4018/978-1-7998-7023-4.ch008

Aydin, A., Akyol, E., Gungor, M., Kaya, A., & Tasdelen, S. (2018). Geophysical Surveys in Engineering Geology Investigations With Field Examples. In N. Ceryan (Ed.), *Handbook of Research on Trends and Digital Advances in Engineering Geology* (pp. 257–280). Hershey, PA: IGI Global. doi:10.4018/978-1-5225-2709-1.ch007

Ayoobkhan, M. U. D., Y., A., J., Easwaran, B., & R., T. (2021). Smart Connected Digital Products and IoT Platform With the Digital Twin. In P. Vasant, G. Weber, & W. Punurai (Ed.), Research Advancements in Smart Technology, Optimization, and Renewable Energy (pp. 330-350). IGI Global. https://doi.org/ doi:10.4018/978-1-7998-3970-5.ch016

Baeza Moyano, D., & González Lezcano, R. A. (2021). The Importance of Light in Our Lives: Towards New Lighting in Schools. In R. González-Lezcano (Ed.), *Advancements in Sustainable Architecture and Energy Efficiency* (pp. 239–256). IGI Global. https://doi.org/10.4018/978-1-7998-7023-4.ch011

Bagdadee, A. H. (2021). A Brief Assessment of the Energy Sector of Bangladesh. *International Journal of Energy Optimization and Engineering*, *10*(1), 36–55. doi:10.4018/IJEOE.2021010103

Baklezos, A. T., & Hadjigeorgiou, N. G. (2021). Magnetic Sensors for Space Applications and Magnetic Cleanliness Considerations. In C. Nikolopoulos (Ed.), *Recent Trends on Electromagnetic Environmental Effects for Aeronautics and Space Applications* (pp. 147–185). IGI Global. https://doi.org/10.4018/978-1-7998-4879-0.ch006

Bas, T. G. (2017). Nutraceutical Industry with the Collaboration of Biotechnology and Nutrigenomics Engineering: The Significance of Intellectual Property in the Entrepreneurship and Scientific Research Ecosystems. In T. Bas & J. Zhao (Eds.), *Comparative Approaches to Biotechnology Development and Use in Developed and Emerging Nations* (pp. 1–17). Hershey, PA: IGI Global. doi:10.4018/978-1-5225-1040-6.ch001

Bazeer Ahamed, B., & Periakaruppan, S. (2021). Taxonomy of Influence Maximization Techniques in Unknown Social Networks. In P. Vasant, G. Weber, & W. Punurai (Eds.), *Research Advancements in Smart Technology, Optimization, and Renewable Energy* (pp. 351-363). IGI Global. https://doi.org/10.4018/978-1-7998-3970-5.ch017

Beale, R., & André, J. (2017). *Design Solutions and Innovations in Temporary Structures*. Hershey, PA: IGI Global. doi:10.4018/978-1-5225-2199-0

Behnam, B. (2017). Simulating Post-Earthquake Fire Loading in Conventional RC Structures. In P. Samui, S. Chakraborty, & D. Kim (Eds.), *Modeling and Simulation Techniques in Structural Engineering* (pp. 425–444). Hershey, PA: IGI Global. doi:10.4018/978-1-5225-0588-4.ch015

Ben Hamida, I., Salah, S. B., Msahli, F., & Mimouni, M. F. (2018). Distribution Network Reconfiguration Using SPEA2 for Power Loss Minimization and Reliability Improvement. *International Journal of Energy Optimization and Engineering*, *7*(1), 50–65. doi:10.4018/IJEOE.2018010103

Bentarzi, H. (2021). Fault Tree-Based Root Cause Analysis Used to Study Mal-Operation of a Protective Relay in a Smart Grid. In A. Recioui & H. Bentarzi (Eds.), *Optimizing and Measuring Smart Grid Operation and Control* (pp. 289–308). IGI Global. https://doi.org/10.4018/978-1-7998-4027-5.ch012

Beysens, D. A., Garrabos, Y., & Zappoli, B. (2021). Thermal Effects in Near-Critical Fluids: Piston Effect and Related Phenomena. In L. Chen (Ed.), *Handbook of Research on Advancements in Supercritical Fluids Applications for Sustainable Energy Systems* (pp. 1–31). IGI Global. https://doi.org/10.4018/978-1-7998-5796-9.ch001

Bhaskar, S. V., & Kudal, H. N. (2017). Effect of TiCN and AlCrN Coating on Tribological Behaviour of Plasma-nitrided AISI 4140 Steel. *International Journal of Surface Engineering and Interdisciplinary Materials Science*, *5*(2), 1–17. doi:10.4018/IJSEIMS.2017070101

Bhuyan, D. (2018). Designing of a Twin Tube Shock Absorber: A Study in Reverse Engineering. In K. Kumar & J. Davim (Eds.), *Design and Optimization of Mechanical Engineering Products* (pp. 83–104). Hershey, PA: IGI Global. doi:10.4018/978-1-5225-3401-3.ch005

Blumberg, G. (2021). Blockchains for Use in Construction and Engineering Projects. In J. Underwood & M. Shelbourn (Eds.), *Handbook of Research on Driving Transformational Change in the Digital Built Environment* (pp. 179–208). IGI Global. https://doi.org/10.4018/978-1-7998-6600-8.ch008

Bolboaca, A. M. (2021). Considerations Regarding the Use of Fuel Cells in Combined Heat and Power for Stationary Applications. In G. Badea, R. Felseghi, & I. Aşchilean (Eds.), *Hydrogen Fuel Cell Technology for Stationary Applications* (pp. 239–275). IGI Global. https://doi.org/10.4018/978-1-7998-4945-2.ch010

Burstein, L. (2021). Simulation Tool for Cable Design. In L. Burstein (Ed.), *Handbook of Research on Advancements in Manufacturing, Materials, and Mechanical Engineering* (pp. 54–74). IGI Global. https://doi.org/10.4018/978-1-7998-4939-1.ch003

Calderon, F. A., Giolo, E. G., Frau, C. D., Rengel, M. G., Rodriguez, H., Tornello, M., ... Gallucci, R. (2018). Seismic Microzonation and Site Effects Detection Through Microtremors Measures: A Review. In N. Ceryan (Ed.), *Handbook of Research on Trends and Digital Advances in Engineering Geology* (pp. 326–349). Hershey, PA: IGI Global. doi:10.4018/978-1-5225-2709-1.ch009

Ceryan, N., & Can, N. K. (2018). Prediction of The Uniaxial Compressive Strength of Rocks Materials. In N. Ceryan (Ed.), *Handbook of Research on Trends and Digital Advances in Engineering Geology* (pp. 31–96). Hershey, PA: IGI Global. doi:10.4018/978-1-5225-2709-1.ch002

Ceryan, S. (2018). Weathering Indices Used in Evaluation of the Weathering State of Rock Material. In N. Ceryan (Ed.), *Handbook of Research on Trends and Digital Advances in Engineering Geology* (pp. 132–186). Hershey, PA: IGI Global. doi:10.4018/978-1-5225-2709-1.ch004

Chen, H., Padilla, R. V., & Besarati, S. (2017). Supercritical Fluids and Their Applications in Power Generation. In L. Chen & Y. Iwamoto (Eds.), *Advanced Applications of Supercritical Fluids in Energy Systems* (pp. 369–402). Hershey, PA: IGI Global. doi:10.4018/978-1-5225-2047-4.ch012

Chen, H., Padilla, R. V., & Besarati, S. (2021). Supercritical Fluids and Their Applications in Power Generation. In L. Chen (Ed.), *Handbook of Research on Advancements in Supercritical Fluids Applications for Sustainable Energy Systems* (pp. 566–599). IGI Global. https://doi.org/10.4018/978-1-7998-5796-9.ch016

Chen, L. (2017). Principles, Experiments, and Numerical Studies of Supercritical Fluid Natural Circulation System. In L. Chen & Y. Iwamoto (Eds.), *Advanced Applications of Supercritical Fluids in Energy Systems* (pp. 136–187). Hershey, PA: IGI Global. doi:10.4018/978-1-5225-2047-4.ch005

Chen, L. (2021). Principles, Experiments, and Numerical Studies of Supercritical Fluid Natural Circulation System. In L. Chen (Ed.), *Handbook of Research on Advancements in Supercritical Fluids Applications for Sustainable Energy Systems* (pp. 219–269). IGI Global. https://doi.org/10.4018/978-1-7998-5796-9.ch007

Chiba, Y., Marif, Y., Henini, N., & Tlemcani, A. (2021). Modeling of Magnetic Refrigeration Device by Using Artificial Neural Networks Approach. *International Journal of Energy Optimization and Engineering*, *10*(4), 68–76. https://doi.org/10.4018/IJEOE.2021100105

Clementi, F., Di Sciascio, G., Di Sciascio, S., & Lenci, S. (2017). Influence of the Shear-Bending Interaction on the Global Capacity of Reinforced Concrete Frames: A Brief Overview of the New Perspectives. In V. Plevris, G. Kremmyda, & Y. Fahjan (Eds.), *Performance-Based Seismic Design of Concrete Structures and Infrastructures* (pp. 84–111). Hershey, PA: IGI Global. doi:10.4018/978-1-5225-2089-4.ch004

Codinhoto, R., Fialho, B. C., Pinti, L., & Fabricio, M. M. (2021). BIM and IoT for Facilities Management: Understanding Key Maintenance Issues. In J. Underwood & M. Shelbourn (Eds.), *Handbook of Research on Driving Transformational Change in the Digital Built Environment* (pp. 209–231). IGI Global. doi:10.4018/978-1-7998-6600-8.ch009

Cortés-Polo, D., Calle-Cancho, J., Carmona-Murillo, J., & González-Sánchez, J. (2017). Future Trends in Mobile-Fixed Integration for Next Generation Networks: Classification and Analysis. *International Journal of Vehicular Telematics and Infotainment Systems*, *1*(1), 33–53. doi:10.4018/IJVTIS.2017010103

Costa, H. G., Sheremetieff, F. H., & Araújo, E. A. (2022). Influence of Game-Based Methods in Developing Engineering Competences. In A. Alves & N. van Hattum-Janssen (Eds.), *Training Engineering Students for Modern Technological Advancement* (pp. 69–88). IGI Global. https://doi.org/10.4018/978-1-7998-8816-1.ch004

Cui, X., Zeng, S., Li, Z., Zheng, Q., Yu, X., & Han, B. (2018). Advanced Composites for Civil Engineering Infrastructures. In K. Kumar & J. Davim (Eds.), *Composites and Advanced Materials for Industrial Applications* (pp. 212–248). Hershey, PA: IGI Global. doi:10.4018/978-1-5225-5216-1.ch010

Dalgıç, S., & Kuşku, İ. (2018). Geological and Geotechnical Investigations in Tunneling. In N. Ceryan (Ed.), *Handbook of Research on Trends and Digital Advances in Engineering Geology* (pp. 482–529). Hershey, PA: IGI Global. doi:10.4018/978-1-5225-2709-1.ch014

Dang, C., & Hihara, E. (2021). Study on Cooling Heat Transfer of Supercritical Carbon Dioxide Applied to Transcritical Carbon Dioxide Heat Pump. In L. Chen (Ed.), *Handbook of Research on Advancements in Supercritical Fluids Applications for Sustainable Energy Systems* (pp. 451–493). IGI Global. https://doi.org/10.4018/978-1-7998-5796-9.ch013

Daus, Y., Kharchenko, V., & Yudaev, I. (2021). Research of Solar Energy Potential of Photovoltaic Installations on Enclosing Structures of Buildings. *International Journal of Energy Optimization and Engineering, 10*(4), 18–34. https://doi.org/10.4018/IJEOE.2021100102

Daus, Y., Kharchenko, V., & Yudaev, I. (2021). Optimizing Layout of Distributed Generation Sources of Power Supply System of Agricultural Object. *International Journal of Energy Optimization and Engineering, 10*(3), 70–84. https://doi.org/10.4018/IJEOE.2021070104

de la Varga, D., Soto, M., Arias, C. A., van Oirschot, D., Kilian, R., Pascual, A., & Álvarez, J. A. (2017). Constructed Wetlands for Industrial Wastewater Treatment and Removal of Nutrients. In Á. Val del Río, J. Campos Gómez, & A. Mosquera Corral (Eds.), *Technologies for the Treatment and Recovery of Nutrients from Industrial Wastewater* (pp. 202–230). Hershey, PA: IGI Global. doi:10.4018/978-1-5225-1037-6.ch008

Deb, S., Ammar, E. A., AlRajhi, H., Alsaidan, I., & Shariff, S. M. (2022). V2G Pilot Projects: Review and Lessons Learnt. In M. Alam, R. Pillai, & N. Murugesan (Eds.), *Developing Charging Infrastructure and Technologies for Electric Vehicles* (pp. 252–267). IGI Global. https://doi.org/10.4018/978-1-7998-6858-3.ch014

Dekhandji, F. Z., & Rais, M. C. (2021). A Comparative Study of Power Quality Monitoring Using Various Techniques. In A. Recioui & H. Bentarzi (Eds.), *Optimizing and Measuring Smart Grid Operation and Control* (pp. 259–288). IGI Global. https://doi.org/10.4018/978-1-7998-4027-5.ch011

Deperlioglu, O. (2018). Intelligent Techniques Inspired by Nature and Used in Biomedical Engineering. In U. Kose, G. Guraksin, & O. Deperlioglu (Eds.), *Nature-Inspired Intelligent Techniques for Solving Biomedical Engineering Problems* (pp. 51–77). Hershey, PA: IGI Global. doi:10.4018/978-1-5225-4769-3.ch003

Dhurpate, P. R., & Tang, H. (2021). Quantitative Analysis of the Impact of Inter-Line Conveyor Capacity for Throughput of Manufacturing Systems. *International Journal of Manufacturing, Materials, and Mechanical Engineering, 11*(1), 1–17. https://doi.org/10.4018/IJMMME.2021010101

Dinkar, S., & Deep, K. (2021). A Survey of Recent Variants and Applications of Antlion Optimizer. *International Journal of Energy Optimization and Engineering, 10*(2), 48–73. doi:10.4018/IJEOE.2021040103

Dixit, A. (2018). Application of Silica-Gel-Reinforced Aluminium Composite on the Piston of Internal Combustion Engine: Comparative Study of Silica-Gel-Reinforced Aluminium Composite Piston With Aluminium Alloy Piston. In K. Kumar & J. Davim (Eds.), *Composites and Advanced Materials for Industrial Applications* (pp. 63–98). Hershey, PA: IGI Global. doi:10.4018/978-1-5225-5216-1.ch004

Drabecki, M. P., & Kułak, K. B. (2021). Global Pandemics on European Electrical Energy Markets: Lessons Learned From the COVID-19 Outbreak. *International Journal of Energy Optimization and Engineering, 10*(3), 24–46. https://doi.org/10.4018/IJEOE.2021070102

Dutta, M. M. (2021). Nanomaterials for Food and Agriculture. In M. Bhat, I. Wani, & S. Ashraf (Eds.), *Applications of Nanomaterials in Agriculture, Food Science, and Medicine* (pp. 75–97). IGI Global. doi:10.4018/978-1-7998-5563-7.ch004

Dutta, M. M., & Goswami, M. (2021). Coating Materials: Nano-Materials. In S. Roy & G. Bose (Eds.), *Advanced Surface Coating Techniques for Modern Industrial Applications* (pp. 1–30). IGI Global. doi:10.4018/978-1-7998-4870-7.ch001

Elsayed, A. M., Dakkama, H. J., Mahmoud, S., Al-Dadah, R., & Kaialy, W. (2017). Sustainable Cooling Research Using Activated Carbon Adsorbents and Their Environmental Impact. In T. Kobayashi (Ed.), *Applied Environmental Materials Science for Sustainability* (pp. 186–221). Hershey, PA: IGI Global. doi:10.4018/978-1-5225-1971-3.ch009

Ercanoglu, M., & Sonmez, H. (2018). General Trends and New Perspectives on Landslide Mapping and Assessment Methods. In N. Ceryan (Ed.), *Handbook of Research on Trends and Digital Advances in Engineering Geology* (pp. 350–379). Hershey, PA: IGI Global. doi:10.4018/978-1-5225-2709-1.ch010

Faroz, S. A., Pujari, N. N., Rastogi, R., & Ghosh, S. (2017). Risk Analysis of Structural Engineering Systems Using Bayesian Inference. In P. Samui, S. Chakraborty, & D. Kim (Eds.), *Modeling and Simulation Techniques in Structural Engineering* (pp. 390–424). Hershey, PA: IGI Global. doi:10.4018/978-1-5225-0588-4.ch014

Fekik, A., Hamida, M. L., Denoun, H., Azar, A. T., Kamal, N. A., Vaidyanathan, S., Bousbaine, A., & Benamrouche, N. (2022). Multilevel Inverter for Hybrid Fuel Cell/PV Energy Conversion System. In A. Fekik & N. Benamrouche (Eds.), *Modeling and Control of Static Converters for Hybrid Storage Systems* (pp. 233–270). IGI Global. https://doi.org/10.4018/978-1-7998-7447-8.ch009

Fekik, A., Hamida, M. L., Houassine, H., Azar, A. T., Kamal, N. A., Denoun, H., Vaidyanathan, S., & Sambas, A. (2022). Power Quality Improvement for Grid-Connected Photovoltaic Panels Using Direct Power Control. In A. Fekik & N. Benamrouche (Eds.), *Modeling and Control of Static Converters for Hybrid Storage Systems* (pp. 107–142). IGI Global. https://doi.org/10.4018/978-1-7998-7447-8.ch005

Fernando, P. R., Hamigah, T., Disne, S., Wickramasingha, G. G., & Sutharshan, A. (2018). The Evaluation of Engineering Properties of Low Cost Concrete Blocks by Partial Doping of Sand with Sawdust: Low Cost Sawdust Concrete Block. *International Journal of Strategic Engineering*, *1*(2), 26–42. doi:10.4018/IJoSE.2018070103

Ferro, G., Minciardi, R., Parodi, L., & Robba, M. (2022). Optimal Charging Management of Microgrid-Integrated Electric Vehicles. In M. Alam, R. Pillai, & N. Murugesan (Eds.), *Developing Charging Infrastructure and Technologies for Electric Vehicles* (pp. 133–155). IGI Global. https://doi.org/10.4018/978-1-7998-6858-3.ch007

Flumerfelt, S., & Green, C. (2022). Graduate Lean Leadership Education: A Case Study of a Program. In A. Alves & N. van Hattum-Janssen (Eds.), *Training Engineering Students for Modern Technological Advancement* (pp. 202–224). IGI Global. https://doi.org/10.4018/978-1-7998-8816-1.ch010

Galli, B. J. (2021). Implications of Economic Decision Making to the Project Manager. *International Journal of Strategic Engineering*, *4*(1), 19–32. https://doi.org/10.4018/IJoSE.2021010102

Gento, A. M., Pimentel, C., & Pascual, J. A. (2022). Teaching Circular Economy and Lean Management in a Learning Factory. In A. Alves & N. van Hattum-Janssen (Eds.), *Training Engineering Students for Modern Technological Advancement* (pp. 183–201). IGI Global. https://doi.org/10.4018/978-1-7998-8816-1.ch009

Ghosh, S., Mitra, S., Ghosh, S., & Chakraborty, S. (2017). Seismic Reliability Analysis in the Framework of Metamodelling Based Monte Carlo Simulation. In P. Samui, S. Chakraborty, & D. Kim (Eds.), *Modeling and Simulation Techniques in Structural Engineering* (pp. 192–208). Hershey, PA: IGI Global. doi:10.4018/978-1-5225-0588-4.ch006

Gil, M., & Otero, B. (2017). Learning Engineering Skills through Creativity and Collaboration: A Game-Based Proposal. In R. Alexandre Peixoto de Queirós & M. Pinto (Eds.), *Gamification-Based E-Learning Strategies for Computer Programming Education* (pp. 14–29). Hershey, PA: IGI Global. doi:10.4018/978-1-5225-1034-5.ch002

Gill, J., Ayre, M., & Mills, J. (2017). Revisioning the Engineering Profession: How to Make It Happen! In M. Gray & K. Thomas (Eds.), *Strategies for Increasing Diversity in Engineering Majors and Careers* (pp. 156–175). Hershey, PA: IGI Global. doi:10.4018/978-1-5225-2212-6.ch008

Godzhaev, Z., Senkevich, S., Kuzmin, V., & Melikov, I. (2021). Use of the Neural Network Controller of Sprung Mass to Reduce Vibrations From Road Irregularities. In P. Vasant, G. Weber, & W. Punurai (Ed.), *Research Advancements in Smart Technology, Optimization, and Renewable Energy* (pp. 69-87). IGI Global. https://doi.org/10.4018/978-1-7998-3970-5.ch005

Gomes de Gusmão, C. M. (2022). Digital Competencies and Transformation in Higher Education: Upskilling With Extension Actions. In A. Alves & N. van Hattum-Janssen (Eds.), *Training Engineering Students for Modern Technological Advancement* (pp. 313–328). IGI Global. https://doi.org/10.4018/978-1-7998-8816-1.ch015A

Goyal, N., Ram, M., & Kumar, P. (2017). Welding Process under Fault Coverage Approach for Reliability and MTTF. In M. Ram & J. Davim (Eds.), *Mathematical Concepts and Applications in Mechanical Engineering and Mechatronics* (pp. 222–245). Hershey, PA: IGI Global. doi:10.4018/978-1-5225-1639-2.ch011

Gray, M., & Lundy, C. (2017). Engineering Study Abroad: High Impact Strategy for Increasing Access. In M. Gray & K. Thomas (Eds.), *Strategies for Increasing Diversity in Engineering Majors and Careers* (pp. 42–59). Hershey, PA: IGI Global. doi:10.4018/978-1-5225-2212-6.ch003

Güler, O., & Varol, T. (2021). Fabrication of Functionally Graded Metal and Ceramic Powders Synthesized by Electroless Deposition. In S. Roy & G. Bose (Eds.), *Advanced Surface Coating Techniques for Modern Industrial Applications* (pp. 150–187). IGI Global. https://doi.org/10.4018/978-1-7998-4870-7.ch007

Guraksin, G. E. (2018). Internet of Things and Nature-Inspired Intelligent Techniques for the Future of Biomedical Engineering. In U. Kose, G. Guraksin, & O. Deperlioglu (Eds.), *Nature-Inspired Intelligent Techniques for Solving Biomedical Engineering Problems* (pp. 263–282). Hershey, PA: IGI Global. doi:10.4018/978-1-5225-4769-3.ch013

Hamida, M. L., Fekik, A., Denoun, H., Ardjal, A., & Bokhtache, A. A. (2022). Flying Capacitor Inverter Integration in a Renewable Energy System. In A. Fekik & N. Benamrouche (Eds.), *Modeling and Control of Static Converters for Hybrid Storage Systems* (pp. 287–306). IGI Global. https://doi.org/10.4018/978-1-7998-7447-8.ch011

Hasegawa, N., & Takahashi, Y. (2021). Control of Soap Bubble Ejection Robot Using Facial Expressions. *International Journal of Manufacturing, Materials, and Mechanical Engineering, 11*(2), 1–16. https://doi.org/10.4018/IJMMME.2021040101

Hejazi, T., & Akbari, L. (2017). A Multiresponse Optimization Model for Statistical Design of Processes with Discrete Variables. In M. Ram & J. Davim (Eds.), *Mathematical Concepts and Applications in Mechanical Engineering and Mechatronics* (pp. 17–37). Hershey, PA: IGI Global. doi:10.4018/978-1-5225-1639-2.ch002

Hejazi, T., & Hejazi, A. (2017). Monte Carlo Simulation for Reliability-Based Design of Automotive Complex Subsystems. In M. Ram & J. Davim (Eds.), *Mathematical Concepts and Applications in Mechanical Engineering and Mechatronics* (pp. 177–200). Hershey, PA: IGI Global. doi:10.4018/978-1-5225-1639-2.ch009

Hejazi, T., & Poursabbagh, H. (2017). Reliability Analysis of Engineering Systems: An Accelerated Life Testing for Boiler Tubes. In M. Ram & J. Davim (Eds.), *Mathematical Concepts and Applications in Mechanical Engineering and Mechatronics* (pp. 154–176). Hershey, PA: IGI Global. doi:10.4018/978-1-5225-1639-2.ch008

Henao, J., Poblano-Salas, C. A., Vargas, F., Giraldo-Betancur, A. L., Corona-Castuera, J., & Sotelo-Mazón, O. (2021). Principles and Applications of Thermal Spray Coatings. In S. Roy & G. Bose (Eds.), *Advanced Surface Coating Techniques for Modern Industrial Applications* (pp. 31–70). IGI Global. https://doi.org/10.4018/978-1-7998-4870-7.ch002

Henao, J., & Sotelo, O. (2018). Surface Engineering at High Temperature: Thermal Cycling and Corrosion Resistance. In A. Pakseresht (Ed.), *Production, Properties, and Applications of High Temperature Coatings* (pp. 131–159). Hershey, PA: IGI Global. doi:10.4018/978-1-5225-4194-3.ch006

Hrnčič, M. K., Cör, D., & Knez, Ž. (2021). Supercritical Fluids as a Tool for Green Energy and Chemicals. In L. Chen (Ed.), *Handbook of Research on Advancements in Supercritical Fluids Applications for Sustainable Energy Systems* (pp. 761–791). IGI Global. doi:10.4018/978-1-7998-5796-9.ch021

Ibrahim, O., Erdem, S., & Gurbuz, E. (2021). Studying Physical and Chemical Properties of Graphene Oxide and Reduced Graphene Oxide and Their Applications in Sustainable Building Materials. In R. González-Lezcano (Ed.), *Advancements in Sustainable Architecture and Energy Efficiency* (pp. 221–238). IGI Global. https://doi.org/10.4018/978-1-7998-7023-4.ch010

Ihianle, I. K., Islam, S., Naeem, U., & Ebenuwa, S. H. (2021). Exploiting Patterns of Object Use for Human Activity Recognition. In A. Nwajana & I. Ihianle (Eds.), *Handbook of Research on 5G Networks and Advancements in Computing, Electronics, and Electrical Engineering* (pp. 382–401). IGI Global. https://doi.org/10.4018/978-1-7998-6992-4.ch015

Ijemaru, G. K., Ngharamike, E. T., Oleka, E. U., & Nwajana, A. O. (2021). An Energy-Efficient Model for Opportunistic Data Collection in IoV-Enabled SC Waste Management. In A. Nwajana & I. Ihianle (Eds.), *Handbook of Research on 5G Networks and Advancements in Computing, Electronics, and Electrical Engineering* (pp. 1–19). IGI Global. https://doi.org/10.4018/978-1-7998-6992-4.ch001

Ilori, O. O., Adetan, D. A., & Umoru, L. E. (2017). Effect of Cutting Parameters on the Surface Residual Stress of Face-Milled Pearlitic Ductile Iron. *International Journal of Materials Forming and Machining Processes*, *4*(1), 38–52. doi:10.4018/IJMFMP.2017010103

Imam, M. H., Tasadduq, I. A., Ahmad, A., Aldosari, F., & Khan, H. (2017). Automated Generation of Course Improvement Plans Using Expert System. *International Journal of Quality Assurance in Engineering and Technology Education*, *6*(1), 1–12. doi:10.4018/IJQAETE.2017010101

Injeti, S. K., & Kumar, T. V. (2018). A WDO Framework for Optimal Deployment of DGs and DSCs in a Radial Distribution System Under Daily Load Pattern to Improve Techno-Economic Benefits. *International Journal of Energy Optimization and Engineering*, *7*(2), 1–38. doi:10.4018/IJEOE.2018040101

Ishii, N., Anami, K., & Knisely, C. W. (2018). *Dynamic Stability of Hydraulic Gates and Engineering for Flood Prevention*. Hershey, PA: IGI Global. doi:10.4018/978-1-5225-3079-4

Iwamoto, Y., & Yamaguchi, H. (2021). Application of Supercritical Carbon Dioxide for Solar Water Heater. In L. Chen (Ed.), *Handbook of Research on Advancements in Supercritical Fluids Applications for Sustainable Energy Systems* (pp. 370–387). IGI Global. https://doi.org/10.4018/978-1-7998-5796-9.ch010

Jayapalan, S. (2018). A Review of Chemical Treatments on Natural Fibers-Based Hybrid Composites for Engineering Applications. In K. Kumar & J. Davim (Eds.), *Composites and Advanced Materials for Industrial Applications* (pp. 16–37). Hershey, PA: IGI Global. doi:10.4018/978-1-5225-5216-1.ch002

Kapetanakis, T. N., Vardiambasis, I. O., Ioannidou, M. P., & Konstantaras, A. I. (2021). Modeling Antenna Radiation Using Artificial Intelligence Techniques: The Case of a Circular Loop Antenna. In C. Nikolopoulos (Ed.), *Recent Trends on Electromagnetic Environmental Effects for Aeronautics and Space Applications* (pp. 186–225). IGI Global. https://doi.org/10.4018/978-1-7998-4879-0.ch007

Karkalos, N. E., Markopoulos, A. P., & Dossis, M. F. (2017). Optimal Model Parameters of Inverse Kinematics Solution of a 3R Robotic Manipulator Using ANN Models. *International Journal of Manufacturing, Materials, and Mechanical Engineering, 7*(3), 20–40. doi:10.4018/IJMMME.2017070102

Kelly, M., Costello, M., Nicholson, G., & O'Connor, J. (2021). The Evolving Integration of BIM Into Built Environment Programmes in a Higher Education Institute. In J. Underwood & M. Shelbourn (Eds.), *Handbook of Research on Driving Transformational Change in the Digital Built Environment* (pp. 294–326). IGI Global. https://doi.org/10.4018/978-1-7998-6600-8.ch012

Kesimal, A., Karaman, K., Cihangir, F., & Ercikdi, B. (2018). Excavatability Assessment of Rock Masses for Geotechnical Studies. In N. Ceryan (Ed.), *Handbook of Research on Trends and Digital Advances in Engineering Geology* (pp. 231–256). Hershey, PA: IGI Global. doi:10.4018/978-1-5225-2709-1.ch006

Knoflacher, H. (2017). The Role of Engineers and Their Tools in the Transport Sector after Paradigm Change: From Assumptions and Extrapolations to Science. In H. Knoflacher & E. Ocalir-Akunal (Eds.), *Engineering Tools and Solutions for Sustainable Transportation Planning* (pp. 1–29). Hershey, PA: IGI Global. doi:10.4018/978-1-5225-2116-7.ch001

Kose, U. (2018). Towards an Intelligent Biomedical Engineering With Nature-Inspired Artificial Intelligence Techniques. In U. Kose, G. Guraksin, & O. Deperlioglu (Eds.), *Nature-Inspired Intelligent Techniques for Solving Biomedical Engineering Problems* (pp. 1–26). Hershey, PA: IGI Global. doi:10.4018/978-1-5225-4769-3.ch001

Kostić, S. (2018). A Review on Enhanced Stability Analyses of Soil Slopes Using Statistical Design. In N. Ceryan (Ed.), *Handbook of Research on Trends and Digital Advances in Engineering Geology* (pp. 446–481). Hershey, PA: IGI Global. doi:10.4018/978-1-5225-2709-1.ch013

Kumar, A., Patil, P. P., & Prajapati, Y. K. (2018). *Advanced Numerical Simulations in Mechanical Engineering*. Hershey, PA: IGI Global. doi:10.4018/978-1-5225-3722-9

Kumar, G. R., Rajyalakshmi, G., & Manupati, V. K. (2017). Surface Micro Patterning of Aluminium Reinforced Composite through Laser Peening. *International Journal of Manufacturing, Materials, and Mechanical Engineering*, *7*(4), 15–27. doi:10.4018/IJMMME.2017100102

Kumar, N., Basu, D. N., & Chen, L. (2021). Effect of Flow Acceleration and Buoyancy on Thermalhydraulics of sCO2 in Mini/Micro-Channel. In L. Chen (Ed.), *Handbook of Research on Advancements in Supercritical Fluids Applications for Sustainable Energy Systems* (pp. 161–182). IGI Global. doi:10.4018/978-1-7998-5796-9.ch005

Kumari, N., & Kumar, K. (2018). Fabrication of Orthotic Calipers With Epoxy-Based Green Composite. In K. Kumar & J. Davim (Eds.), *Composites and Advanced Materials for Industrial Applications* (pp. 157–176). Hershey, PA: IGI Global. doi:10.4018/978-1-5225-5216-1.ch008

Kuppusamy, R. R. (2018). Development of Aerospace Composite Structures Through Vacuum-Enhanced Resin Transfer Moulding Technology (VERTMTy): Vacuum-Enhanced Resin Transfer Moulding. In K. Kumar & J. Davim (Eds.), *Composites and Advanced Materials for Industrial Applications* (pp. 99–111). Hershey, PA: IGI Global. doi:10.4018/978-1-5225-5216-1.ch005

Kurganov, V. A., Zeigarnik, Y. A., & Maslakova, I. V. (2021). Normal and Deteriorated Heat Transfer Under Heating Turbulent Supercritical Pressure Coolants Flows in Round Tubes. In L. Chen (Ed.), *Handbook of Research on Advancements in Supercritical Fluids Applications for Sustainable Energy Systems* (pp. 494–532). IGI Global. https://doi.org/10.4018/978-1-7998-5796-9.ch014

Li, H., & Zhang, Y. (2021). Heat Transfer and Fluid Flow Modeling for Supercritical Fluids in Advanced Energy Systems. In L. Chen (Ed.), *Handbook of Research on Advancements in Supercritical Fluids Applications for Sustainable Energy Systems* (pp. 388–422). IGI Global. https://doi.org/10.4018/978-1-7998-5796-9.ch011

Loy, J., Howell, S., & Cooper, R. (2017). Engineering Teams: Supporting Diversity in Engineering Education. In M. Gray & K. Thomas (Eds.), *Strategies for Increasing Diversity in Engineering Majors and Careers* (pp. 106–129). Hershey, PA: IGI Global. doi:10.4018/978-1-5225-2212-6.ch006

Macher, G., Armengaud, E., Kreiner, C., Brenner, E., Schmittner, C., Ma, Z., ... Krammer, M. (2018). Integration of Security in the Development Lifecycle of Dependable Automotive CPS. In N. Druml, A. Genser, A. Krieg, M. Menghin, & A. Hoeller (Eds.), *Solutions for Cyber-Physical Systems Ubiquity* (pp. 383–423). Hershey, PA: IGI Global. doi:10.4018/978-1-5225-2845-6.ch015

Madhu, M. N., Singh, J. G., Mohan, V., & Ongsakul, W. (2021). Transmission Risk Optimization in Interconnected Systems: Risk-Adjusted Available Transfer Capability. In P. Vasant, G. Weber, & W. Punurai (Ed.), *Research Advancements in Smart Technology, Optimization, and Renewable Energy* (pp. 183-199). IGI Global. https://doi.org/10.4018/978-1-7998-3970-5.ch010

Mahendramani, G., & Lakshmana Swamy, N. (2018). Effect of Weld Groove Area on Distortion of Butt Welded Joints in Submerged Arc Welding. *International Journal of Manufacturing, Materials, and Mechanical Engineering*, *8*(2), 33–44. doi:10.4018/IJMMME.2018040103

Makropoulos, G., Koumaras, H., Setaki, F., Filis, K., Lutz, T., Montowtt, P., Tomaszewski, L., Dybiec, P., & Järvet, T. (2021). 5G and Unmanned Aerial Vehicles (UAVs) Use Cases: Analysis of the Ecosystem, Architecture, and Applications. In A. Nwajana & I. Ihianle (Eds.), *Handbook of Research on 5G Networks and Advancements in Computing, Electronics, and Electrical Engineering* (pp. 36–69). IGI Global. https://doi.org/10.4018/978-1-7998-6992-4.ch003

Meric, E. M., Erdem, S., & Gurbuz, E. (2021). Application of Phase Change Materials in Construction Materials for Thermal Energy Storage Systems in Buildings. In R. González-Lezcano (Ed.), *Advancements in Sustainable Architecture and Energy Efficiency* (pp. 1–20). IGI Global. https://doi.org/10.4018/978-1-7998-7023-4.ch001

Mihret, E. T., & Yitayih, K. A. (2021). Operation of VANET Communications: The Convergence of UAV System With LTE/4G and WAVE Technologies. *International Journal of Smart Vehicles and Smart Transportation*, *4*(1), 29–51. https://doi.org/10.4018/IJSVST.2021010103

Mir, M. A., Bhat, B. A., Sheikh, B. A., Rather, G. A., Mehraj, S., & Mir, W. R. (2021). Nanomedicine in Human Health Therapeutics and Drug Delivery: Nanobiotechnology and Nanobiomedicine. In M. Bhat, I. Wani, & S. Ashraf (Eds.), *Applications of Nanomaterials in Agriculture, Food Science, and Medicine* (pp. 229–251). IGI Global. doi:10.4018/978-1-7998-5563-7.ch013

Mohammadzadeh, S., & Kim, Y. (2017). Nonlinear System Identification of Smart Buildings. In P. Samui, S. Chakraborty, & D. Kim (Eds.), *Modeling and Simulation Techniques in Structural Engineering* (pp. 328–347). Hershey, PA: IGI Global. doi:10.4018/978-1-5225-0588-4.ch011

Molina, G. J., Aktaruzzaman, F., Soloiu, V., & Rahman, M. (2017). Design and Testing of a Jet-Impingement Instrument to Study Surface-Modification Effects by Nanofluids. *International Journal of Surface Engineering and Interdisciplinary Materials Science*, *5*(2), 43–61. doi:10.4018/IJSEIMS.2017070104

Moreno-Rangel, A., & Carrillo, G. (2021). Energy-Efficient Homes: A Heaven for Respiratory Illnesses. In R. González-Lezcano (Ed.), *Advancements in Sustainable Architecture and Energy Efficiency* (pp. 49–71). IGI Global. https://doi.org/10.4018/978-1-7998-7023-4.ch003

Msomi, V., & Jantjies, B. T. (2021). Correlative Analysis Between Tensile Properties and Tool Rotational Speeds of Friction Stir Welded Similar Aluminium Alloy Joints. *International Journal of Surface Engineering and Interdisciplinary Materials Science*, *9*(2), 58–78. https://doi.org/10.4018/IJSEIMS.2021070104

Muigai, M. N., Mwema, F. M., Akinlabi, E. T., & Obiko, J. O. (2021). Surface Engineering of Materials Through Weld-Based Technologies: An Overview. In S. Roy & G. Bose (Eds.), *Advanced Surface Coating Techniques for Modern Industrial Applications* (pp. 247–260). IGI Global. doi:10.4018/978-1-7998-4870-7.ch011

Mukherjee, A., Saeed, R. A., Dutta, S., & Naskar, M. K. (2017). Fault Tracking Framework for Software-Defined Networking (SDN). In C. Singhal & S. De (Eds.), *Resource Allocation in Next-Generation Broadband Wireless Access Networks* (pp. 247–272). Hershey, PA: IGI Global. doi:10.4018/978-1-5225-2023-8.ch011

Mukhopadhyay, A., Barman, T. K., & Sahoo, P. (2018). Electroless Nickel Coatings for High Temperature Applications. In K. Kumar & J. Davim (Eds.), *Composites and Advanced Materials for Industrial Applications* (pp. 297–331). Hershey, PA: IGI Global. doi:10.4018/978-1-5225-5216-1.ch013

Mwema, F. M., & Wambua, J. M. (2022). Machining of Poly Methyl Methacrylate (PMMA) and Other Olymeric Materials: A Review. In K. Kumar, B. Babu, & J. Davim (Eds.), *Handbook of Research on Advancements in the Processing, Characterization, and Application of Lightweight Materials* (pp. 363–379). IGI Global. https://doi.org/10.4018/978-1-7998-7864-3.ch016

Mykhailyshyn, R., Savkiv, V., Boyko, I., Prada, E., & Virgala, I. (2021). Substantiation of Parameters of Friction Elements of Bernoulli Grippers With a Cylindrical Nozzle. *International Journal of Manufacturing, Materials, and Mechanical Engineering, 11*(2), 17–39. https://doi.org/10.4018/IJMMME.2021040102

Náprstek, J., & Fischer, C. (2017). Dynamic Stability and Post-Critical Processes of Slender Auto-Parametric Systems. In V. Plevris, G. Kremmyda, & Y. Fahjan (Eds.), *Performance-Based Seismic Design of Concrete Structures and Infrastructures* (pp. 128–171). Hershey, PA: IGI Global. doi:10.4018/978-1-5225-2089-4.ch006

Nautiyal, L., Shivach, P., & Ram, M. (2018). Optimal Designs by Means of Genetic Algorithms. In M. Ram & J. Davim (Eds.), *Soft Computing Techniques and Applications in Mechanical Engineering* (pp. 151–161). Hershey, PA: IGI Global. doi:10.4018/978-1-5225-3035-0.ch007

Nazir, R. (2017). Advanced Nanomaterials for Water Engineering and Treatment: Nano-Metal Oxides and Their Nanocomposites. In T. Saleh (Ed.), *Advanced Nanomaterials for Water Engineering, Treatment, and Hydraulics* (pp. 84–126). Hershey, PA: IGI Global. doi:10.4018/978-1-5225-2136-5.ch005

Nikolopoulos, C. D. (2021). Recent Advances on Measuring and Modeling ELF-Radiated Emissions for Space Applications. In C. Nikolopoulos (Ed.), *Recent Trends on Electromagnetic Environmental Effects for Aeronautics and Space Applications* (pp. 1–38). IGI Global. https://doi.org/10.4018/978-1-7998-4879-0.ch001

Nogueira, A. F., Ribeiro, J. C., Fernández de Vega, F., & Zenha-Rela, M. A. (2018). Evolutionary Approaches to Test Data Generation for Object-Oriented Software: Overview of Techniques and Tools. In M. Khosrow-Pour, D.B.A. (Ed.), Incorporating Nature-Inspired Paradigms in Computational Applications (pp. 162-194). Hershey, PA: IGI Global. https://doi.org/ doi:10.4018/978-1-5225-5020-4.ch006

Nwajana, A. O., Obi, E. R., Ijemaru, G. K., Oleka, E. U., & Anthony, D. C. (2021). Fundamentals of RF/Microwave Bandpass Filter Design. In A. Nwajana & I. Ihianle (Eds.), *Handbook of Research on 5G Networks and Advancements in Computing, Electronics, and Electrical Engineering* (pp. 149–164). IGI Global. https://doi.org/10.4018/978-1-7998-6992-4.ch005

Ogbodo, E. A. (2021). Comparative Study of Transmission Line Junction vs. Asynchronously Coupled Junction Diplexers. In A. Nwajana & I. Ihianle (Eds.), *Handbook of Research on 5G Networks and Advancements in Computing, Electronics, and Electrical Engineering* (pp. 326–336). IGI Global. https://doi.org/10.4018/978-1-7998-6992-4.ch013

Orosa, J. A., Vergara, D., Fraguela, F., & Masdías-Bonome, A. (2021). Statistical Understanding and Optimization of Building Energy Consumption and Climate Change Consequences. In R. González-Lezcano (Ed.), *Advancements in Sustainable Architecture and Energy Efficiency* (pp. 195–220). IGI Global. https://doi.org/10.4018/978-1-7998-7023-4.ch009

Osho, M. B. (2018). Industrial Enzyme Technology: Potential Applications. In S. Bharati & P. Chaurasia (Eds.), *Research Advancements in Pharmaceutical, Nutritional, and Industrial Enzymology* (pp. 375–394). Hershey, PA: IGI Global. doi:10.4018/978-1-5225-5237-6.ch017

Ouadi, A., & Zitouni, A. (2021). Phasor Measurement Improvement Using Digital Filter in a Smart Grid. In A. Recioui & H. Bentarzi (Eds.), *Optimizing and Measuring Smart Grid Operation and Control* (pp. 100–117). IGI Global. https://doi.org/10.4018/978-1-7998-4027-5.ch005

Padmaja, P., & Marutheswar, G. (2017). Certain Investigation on Secured Data Transmission in Wireless Sensor Networks. *International Journal of Mobile Computing and Multimedia Communications*, *8*(1), 48–61. doi:10.4018/IJMCMC.2017010104

Palmer, S., & Hall, W. (2017). An Evaluation of Group Work in First-Year Engineering Design Education. In R. Tucker (Ed.), *Collaboration and Student Engagement in Design Education* (pp. 145–168). Hershey, PA: IGI Global. doi:10.4018/978-1-5225-0726-0.ch007

Panchenko, V. (2021). Prospects for Energy Supply of the Arctic Zone Objects of Russia Using Frost-Resistant Solar Modules. In P. Vasant, G. Weber, & W. Punurai (Eds.), *Research Advancements in Smart Technology, Optimization, and Renewable Energy* (pp. 149-169). IGI Global. https://doi.org/10.4018/978-1-7998-3970-5.ch008

Panchenko, V. (2021). Photovoltaic Thermal Module With Paraboloid Type Solar Concentrators. *International Journal of Energy Optimization and Engineering*, *10*(2), 1–23. https://doi.org/10.4018/IJEOE.2021040101

Pandey, K., & Datta, S. (2021). Dry Machining of Inconel 825 Superalloys: Performance of Tool Inserts (Carbide, Cermet, and SiAlON). *International Journal of Manufacturing, Materials, and Mechanical Engineering*, *11*(4), 26–39. doi:10.4018/IJMMME.2021100102

Panneer, R. (2017). Effect of Composition of Fibers on Properties of Hybrid Composites. *International Journal of Manufacturing, Materials, and Mechanical Engineering*, *7*(4), 28–43. doi:10.4018/IJMMME.2017100103

Pany, C. (2021). Estimation of Correct Long-Seam Mismatch Using FEA to Compare the Measured Strain in a Non-Destructive Testing of a Pressurant Tank: A Reverse Problem. *International Journal of Smart Vehicles and Smart Transportation*, *4*(1), 16–28. doi:10.4018/IJSVST.2021010102

Paul, S., & Roy, P. (2018). Optimal Design of Power System Stabilizer Using a Novel Evolutionary Algorithm. *International Journal of Energy Optimization and Engineering*, *7*(3), 24–46. doi:10.4018/IJEOE.2018070102

Paul, S., & Roy, P. K. (2021). Oppositional Differential Search Algorithm for the Optimal Tuning of Both Single Input and Dual Input Power System Stabilizer. In P. Vasant, G. Weber, & W. Punurai (Eds.), *Research Advancements in Smart Technology, Optimization, and Renewable Energy* (pp. 256-282). IGI Global. https://doi.org/10.4018/978-1-7998-3970-5.ch013

Pavaloiu, A. (2018). Artificial Intelligence Ethics in Biomedical-Engineering-Oriented Problems. In U. Kose, G. Guraksin, & O. Deperlioglu (Eds.), *Nature-Inspired Intelligent Techniques for Solving Biomedical Engineering Problems* (pp. 219–231). Hershey, PA: IGI Global. doi:10.4018/978-1-5225-4769-3.ch010

Pioro, I., Mahdi, M., & Popov, R. (2017). Application of Supercritical Pressures in Power Engineering. In L. Chen & Y. Iwamoto (Eds.), *Advanced Applications of Supercritical Fluids in Energy Systems* (pp. 404–457). Hershey, PA: IGI Global. doi:10.4018/978-1-5225-2047-4.ch013

Plaksina, T., & Gildin, E. (2017). Rigorous Integrated Evolutionary Workflow for Optimal Exploitation of Unconventional Gas Assets. *International Journal of Energy Optimization and Engineering*, *6*(1), 101–122. doi:10.4018/IJEOE.2017010106

Popat, J., Kakadiya, H., Tak, L., Singh, N. K., Majeed, M. A., & Mahajan, V. (2021). Reliability of Smart Grid Including Cyber Impact: A Case Study. In R. Singh, A. Singh, A. Dwivedi, & P. Nagabhushan (Eds.), *Computational Methodologies for Electrical and Electronics Engineers* (pp. 163–174). IGI Global. https://doi.org/10.4018/978-1-7998-3327-7.ch013

Quiza, R., La Fé-Perdomo, I., Rivas, M., & Ramtahalsing, V. (2021). Triple Bottom Line-Focused Optimization of Oblique Turning Processes Based on Hybrid Modeling: A Study Case on AISI 1045 Steel Turning. In L. Burstein (Ed.), *Handbook of Research on Advancements in Manufacturing, Materials, and Mechanical Engineering* (pp. 215–241). IGI Global. https://doi.org/10.4018/978-1-7998-4939-1.ch010

Rahmani, M. K. (2022). Blockchain Technology: Principles and Algorithms. In S. Khan, M. Syed, R. Hammad, & A. Bushager (Eds.), *Blockchain Technology and Computational Excellence for Society 5.0* (pp. 16–27). IGI Global. https://doi.org/10.4018/978-1-7998-8382-1.ch002

Ramdani, N., & Azibi, M. (2018). Polymer Composite Materials for Microelectronics Packaging Applications: Composites for Microelectronics Packaging. In K. Kumar & J. Davim (Eds.), *Composites and Advanced Materials for Industrial Applications* (pp. 177–211). Hershey, PA: IGI Global. doi:10.4018/978-1-5225-5216-1.ch009

Ramesh, M., Garg, R., & Subrahmanyam, G. V. (2017). Investigation of Influence of Quenching and Annealing on the Plane Fracture Toughness and Brittle to Ductile Transition Temperature of the Zinc Coated Structural Steel Materials. *International Journal of Surface Engineering and Interdisciplinary Materials Science*, *5*(2), 33–42. doi:10.4018/IJSEIMS.2017070103

Robinson, J., & Beneroso, D. (2022). Project-Based Learning in Chemical Engineering: Curriculum and Assessment, Culture and Learning Spaces. In A. Alves & N. van Hattum-Janssen (Eds.), *Training Engineering Students for Modern Technological Advancement* (pp. 1–19). IGI Global. https://doi.org/10.4018/978-1-7998-8816-1.ch001

Rondon, B. (2021). Experimental Characterization of Admittance Meter With Crude Oil Emulsions. *International Journal of Electronics, Communications, and Measurement Engineering*, *10*(2), 51–59. https://doi.org/10.4018/IJECME.2021070104

Rudolf, S., Biryuk, V. V., & Volov, V. (2018). Vortex Effect, Vortex Power: Technology of Vortex Power Engineering. In V. Kharchenko & P. Vasant (Eds.), *Handbook of Research on Renewable Energy and Electric Resources for Sustainable Rural Development* (pp. 500–533). Hershey, PA: IGI Global. doi:10.4018/978-1-5225-3867-7.ch021

Sah, A., Bhadula, S. J., Dumka, A., & Rawat, S. (2018). A Software Engineering Perspective for Development of Enterprise Applications. In A. Elçi (Ed.), *Handbook of Research on Contemporary Perspectives on Web-Based Systems* (pp. 1–23). Hershey, PA: IGI Global. doi:10.4018/978-1-5225-5384-7.ch001

Sahli, Y., Zitouni, B., & Hocine, B. M. (2021). Three-Dimensional Numerical Study of Overheating of Two Intermediate Temperature P-AS-SOFC Geometrical Configurations. In G. Badea, R. Felseghi, & I. Aşchilean (Eds.), *Hydrogen Fuel Cell Technology for Stationary Applications* (pp. 186–222). IGI Global. https://doi.org/10.4018/978-1-7998-4945-2.ch008

Sahoo, P., & Roy, S. (2017). Tribological Behavior of Electroless Ni-P, Ni-P-W and Ni-P-Cu Coatings: A Comparison. *International Journal of Surface Engineering and Interdisciplinary Materials Science*, *5*(1), 1–15. doi:10.4018/IJSEIMS.2017010101

Sahoo, S. (2018). Laminated Composite Hypar Shells as Roofing Units: Static and Dynamic Behavior. In K. Kumar & J. Davim (Eds.), *Composites and Advanced Materials for Industrial Applications* (pp. 249–269). Hershey, PA: IGI Global. doi:10.4018/978-1-5225-5216-1.ch011

Sahu, H., & Hungyo, M. (2018). Introduction to SDN and NFV. In A. Dumka (Ed.), *Innovations in Software-Defined Networking and Network Functions Virtualization* (pp. 1–25). Hershey, PA: IGI Global. doi:10.4018/978-1-5225-3640-6.ch001

Salem, A. M., & Shmelova, T. (2018). Intelligent Expert Decision Support Systems: Methodologies, Applications, and Challenges. In T. Shmelova, Y. Sikirda, N. Rizun, A. Salem, & Y. Kovalyov (Eds.), *Socio-Technical Decision Support in Air Navigation Systems: Emerging Research and Opportunities* (pp. 215–242). Hershey, PA: IGI Global. doi:10.4018/978-1-5225-3108-1.ch007

Samal, M. (2017). FE Analysis and Experimental Investigation of Cracked and Un-Cracked Thin-Walled Tubular Components to Evaluate Mechanical and Fracture Properties. In P. Samui, S. Chakraborty, & D. Kim (Eds.), *Modeling and Simulation Techniques in Structural Engineering* (pp. 266–293). Hershey, PA: IGI Global. doi:10.4018/978-1-5225-0588-4.ch009

Samal, M., & Balakrishnan, K. (2017). Experiments on a Ring Tension Setup and FE Analysis to Evaluate Transverse Mechanical Properties of Tubular Components. In P. Samui, S. Chakraborty, & D. Kim (Eds.), *Modeling and Simulation Techniques in Structural Engineering* (pp. 91–115). Hershey, PA: IGI Global. doi:10.4018/978-1-5225-0588-4.ch004

Samarasinghe, D. A., & Wood, E. (2021). Innovative Digital Technologies. In J. Underwood & M. Shelbourn (Eds.), *Handbook of Research on Driving Transformational Change in the Digital Built Environment* (pp. 142–163). IGI Global. https://doi.org/10.4018/978-1-7998-6600-8.ch006

Sawant, S. (2018). Deep Learning and Biomedical Engineering. In U. Kose, G. Guraksin, & O. Deperlioglu (Eds.), *Nature-Inspired Intelligent Techniques for Solving Biomedical Engineering Problems* (pp. 283–296). Hershey, PA: IGI Global. doi:10.4018/978-1-5225-4769-3.ch014

Schulenberg, T. (2021). Energy Conversion Using the Supercritical Steam Cycle. In L. Chen (Ed.), *Handbook of Research on Advancements in Supercritical Fluids Applications for Sustainable Energy Systems* (pp. 659–681). IGI Global. doi:10.4018/978-1-7998-5796-9.ch018

Sezgin, H., & Berkalp, O. B. (2018). Textile-Reinforced Composites for the Automotive Industry. In K. Kumar & J. Davim (Eds.), *Composites and Advanced Materials for Industrial Applications* (pp. 129–156). Hershey, PA: IGI Global. doi:10.4018/978-1-5225-5216-1.ch007

Shaaban, A. A., & Shehata, O. M. (2021). Combining Response Surface Method and Metaheuristic Algorithms for Optimizing SPIF Process. *International Journal of Manufacturing, Materials, and Mechanical Engineering*, *11*(4), 1–25. https://doi.org/10.4018/IJMMME.2021100101

Shafaati Shemami, M., & Sefid, M. (2022). Implementation and Demonstration of Electric Vehicle-to-Home (V2H) Application: A Case Study. In M. Alam, R. Pillai, & N. Murugesan (Eds.), *Developing Charging Infrastructure and Technologies for Electric Vehicles* (pp. 268–293). IGI Global. https://doi.org/10.4018/978-1-7998-6858-3.ch015

Shah, M. Z., Gazder, U., Bhatti, M. S., & Hussain, M. (2018). Comparative Performance Evaluation of Effects of Modifier in Asphaltic Concrete Mix. *International Journal of Strategic Engineering*, *1*(2), 13–25. doi:10.4018/IJoSE.2018070102

Sharma, N., & Kumar, K. (2018). Fabrication of Porous NiTi Alloy Using Organic Binders. In K. Kumar & J. Davim (Eds.), *Composites and Advanced Materials for Industrial Applications* (pp. 38–62). Hershey, PA: IGI Global. doi:10.4018/978-1-5225-5216-1.ch003

Shivach, P., Nautiyal, L., & Ram, M. (2018). Applying Multi-Objective Optimization Algorithms to Mechanical Engineering. In M. Ram & J. Davim (Eds.), *Soft Computing Techniques and Applications in Mechanical Engineering* (pp. 287–301). Hershey, PA: IGI Global. doi:10.4018/978-1-5225-3035-0.ch014

Shmelova, T. (2018). Stochastic Methods for Estimation and Problem Solving in Engineering: Stochastic Methods of Decision Making in Aviation. In S. Kadry (Ed.), *Stochastic Methods for Estimation and Problem Solving in Engineering* (pp. 139–160). Hershey, PA: IGI Global. doi:10.4018/978-1-5225-5045-7.ch006

Siero González, L. R., & Romo Vázquez, A. (2017). Didactic Sequences Teaching Mathematics for Engineers With Focus on Differential Equations. In M. Ramírez-Montoya (Ed.), *Handbook of Research on Driving STEM Learning With Educational Technologies* (pp. 129–151). Hershey, PA: IGI Global. doi:10.4018/978-1-5225-2026-9.ch007

Sim, M. S., You, K. Y., Esa, F., & Chan, Y. L. (2021). Nanostructured Electromagnetic Metamaterials for Sensing Applications. In M. Bhat, I. Wani, & S. Ashraf (Eds.), *Applications of Nanomaterials in Agriculture, Food Science, and Medicine* (pp. 141–164). IGI Global. https://doi.org/10.4018/978-1-7998-5563-7.ch009

Singh, R., & Dutta, S. (2018). Visible Light Active Nanocomposites for Photocatalytic Applications. In K. Kumar & J. Davim (Eds.), *Composites and Advanced Materials for Industrial Applications* (pp. 270–296). Hershey, PA: IGI Global. doi:10.4018/978-1-5225-5216-1.ch012

Skripov, P. V., Yampol'skiy, A. D., & Rutin, S. B. (2021). High-Power Heat Transfer in Supercritical Fluids: Microscale Times and Sizes. In L. Chen (Ed.), *Handbook of Research on Advancements in Supercritical Fluids Applications for Sustainable Energy Systems* (pp. 424–450). IGI Global. https://doi.org/10.4018/978-1-7998-5796-9.ch012

Sözbilir, H., Özkaymak, Ç., Uzel, B., & Sümer, Ö. (2018). Criteria for Surface Rupture Microzonation of Active Faults for Earthquake Hazards in Urban Areas. In N. Ceryan (Ed.), *Handbook of Research on Trends and Digital Advances in Engineering Geology* (pp. 187–230). Hershey, PA: IGI Global. doi:10.4018/978-1-5225-2709-1.ch005

Stanciu, I. (2018). Stochastic Methods in Microsystems Engineering. In S. Kadry (Ed.), *Stochastic Methods for Estimation and Problem Solving in Engineering* (pp. 161–176). Hershey, PA: IGI Global. doi:10.4018/978-1-5225-5045-7.ch007

Strebkov, D., Nekrasov, A., Trubnikov, V., & Nekrasov, A. (2018). Single-Wire Resonant Electric Power Systems for Renewable-Based Electric Grid. In V. Kharchenko & P. Vasant (Eds.), *Handbook of Research on Renewable Energy and Electric Resources for Sustainable Rural Development* (pp. 449–474). Hershey, PA: IGI Global. doi:10.4018/978-1-5225-3867-7.ch019

Sukhyy, K., Belyanovskaya, E., & Sukhyy, M. (2021). *Basic Principles for Substantiation of Working Pair Choice*. IGI Global. doi:10.4018/978-1-7998-4432-7.ch002

Suri, M. S., & Kaliyaperumal, D. (2022). Extension of Aspiration Level Model for Optimal Planning of Fast Charging Stations. In A. Fekik & N. Benamrouche (Eds.), *Modeling and Control of Static Converters for Hybrid Storage Systems* (pp. 91–106). IGI Global. https://doi.org/10.4018/978-1-7998-7447-8.ch004

Tallet, E., Gledson, B., Rogage, K., Thompson, A., & Wiggett, D. (2021). Digitally-Enabled Design Management. In J. Underwood & M. Shelbourn (Eds.), *Handbook of Research on Driving Transformational Change in the Digital Built Environment* (pp. 63–89). IGI Global. https://doi.org/10.4018/978-1-7998-6600-8.ch003

Terki, A., & Boubertakh, H. (2021). A New Hybrid Binary-Real Coded Cuckoo Search and Tabu Search Algorithm for Solving the Unit-Commitment Problem. *International Journal of Energy Optimization and Engineering, 10*(2), 104–119. https://doi.org/10.4018/IJEOE.2021040105

Tüdeş, Ş., Kumlu, K. B., & Ceryan, S. (2018). Integration Between Urban Planning and Natural Hazards For Resilient City. In N. Ceryan (Ed.), *Handbook of Research on Trends and Digital Advances in Engineering Geology* (pp. 591–630). Hershey, PA: IGI Global. doi:10.4018/978-1-5225-2709-1.ch017

Ulamis, K. (2018). Soil Liquefaction Assessment by Anisotropic Cyclic Triaxial Test. In N. Ceryan (Ed.), *Handbook of Research on Trends and Digital Advances in Engineering Geology* (pp. 631–664). Hershey, PA: IGI Global. doi:10.4018/978-1-5225-2709-1.ch018

Valente, M., & Milani, G. (2017). Seismic Assessment and Retrofitting of an Under-Designed RC Frame Through a Displacement-Based Approach. In V. Plevris, G. Kremmyda, & Y. Fahjan (Eds.), *Performance-Based Seismic Design of Concrete Structures and Infrastructures* (pp. 36–58). Hershey, PA: IGI Global. doi:10.4018/978-1-5225-2089-4.ch002

Vargas-Bernal, R. (2021). Advances in Electromagnetic Environmental Shielding for Aeronautics and Space Applications. In C. Nikolopoulos (Ed.), *Recent Trends on Electromagnetic Environmental Effects for Aeronautics and Space Applications* (pp. 80–96). IGI Global. https://doi.org/10.4018/978-1-7998-4879-0.ch003

Vasant, P. (2018). A General Medical Diagnosis System Formed by Artificial Neural Networks and Swarm Intelligence Techniques. In U. Kose, G. Guraksin, & O. Deperlioglu (Eds.), *Nature-Inspired Intelligent Techniques for Solving Biomedical Engineering Problems* (pp. 130–145). Hershey, PA: IGI Global. doi:10.4018/978-1-5225-4769-3.ch006

Verner, C. M., & Sarwar, D. (2021). Avoiding Project Failure and Achieving Project Success in NHS IT System Projects in the United Kingdom. *International Journal of Strategic Engineering*, *4*(1), 33–54. https://doi.org/10.4018/IJoSE.2021010103

Verrollot, J., Tolonen, A., Harkonen, J., & Haapasalo, H. J. (2018). Challenges and Enablers for Rapid Product Development. *International Journal of Applied Industrial Engineering*, *5*(1), 25–49. doi:10.4018/IJAIE.2018010102

Wan, A. C., Zulu, S. L., & Khosrow-Shahi, F. (2021). Industry Views on BIM for Site Safety in Hong Kong. In J. Underwood & M. Shelbourn (Eds.), *Handbook of Research on Driving Transformational Change in the Digital Built Environment* (pp. 120–140). IGI Global. https://doi.org/10.4018/978-1-7998-6600-8.ch005

Yardimci, A. G., & Karpuz, C. (2018). Fuzzy Rock Mass Rating: Soft-Computing-Aided Preliminary Stability Analysis of Weak Rock Slopes. In N. Ceryan (Ed.), *Handbook of Research on Trends and Digital Advances in Engineering Geology* (pp. 97–131). Hershey, PA: IGI Global. doi:10.4018/978-1-5225-2709-1.ch003

You, K. Y. (2021). Development Electronic Design Automation for RF/Microwave Antenna Using MATLAB GUI. In A. Nwajana & I. Ihianle (Eds.), *Handbook of Research on 5G Networks and Advancements in Computing, Electronics, and Electrical Engineering* (pp. 70–148). IGI Global. https://doi.org/10.4018/978-1-7998-6992-4.ch004

Yousefi, Y., Gratton, P., & Sarwar, D. (2021). Investigating the Opportunities to Improve the Thermal Performance of a Case Study Building in London. *International Journal of Strategic Engineering*, *4*(1), 1–18. https://doi.org/10.4018/IJoSE.2021010101

Zindani, D., & Kumar, K. (2018). Industrial Applications of Polymer Composite Materials. In K. Kumar & J. Davim (Eds.), *Composites and Advanced Materials for Industrial Applications* (pp. 1–15). Hershey, PA: IGI Global. doi:10.4018/978-1-5225-5216-1.ch001

Zindani, D., Maity, S. R., & Bhowmik, S. (2018). A Decision-Making Approach for Material Selection of Polymeric Composite Bumper Beam. In K. Kumar & J. Davim (Eds.), *Composites and Advanced Materials for Industrial Applications* (pp. 112–128). Hershey, PA: IGI Global. doi:10.4018/978-1-5225-5216-1.ch006

About the Contributors

Krishnaraj Nagappan is working as a Associate Professor, School of Computing, SRM Institute of Science and Technology, Kattankulathur, Tamilnadu, India. He is having 13 years of experience in teaching and research, his research areas are Biometrics, Wireless sensor networks, Internet of Things, Medical image processing. He has completed one funded research project supported by DST, India. He is Cisco certified Routing and switching professional. He has published more than 65 articles in reputed international journals and 15 articles presented in international conferences. He is being serving as a Editorial board member of MAT Journal, IRED and Allied Academic Sciences. He has delivered several special lectures in workshops and seminars. He is a professional society member for ISTE, IEI and IAENG.

* * *

Kalpana A.V. is an Assistant Professor.

Shivakumar Hunagund is an experienced Senior research engineer (semiconductor technology development) with Intel corporation, currently working on next gen product development in Microprocessor and Graphics with a focus on failure analysis. His 8 years in research and development work in Quantum computing, Materials/Nuclear Engineering, Photonics and Nanotechnology has taken him across the Europe and US.

Ramprasath M. is an Assistant Professor.

Kamaleswari P. has more than ten years of teaching experience.

Nadana Ravishankar T. is an Assistant Professor.

Index

N

O

P

Q

Recommended Reference Books

IGI Global's reference books are available in three unique pricing formats:
Print Only, E-Book Only, or Print + E-Book.

Order direct through IGI Global's Online Bookstore at **www.igi-global.com** or through your preferred provider.

ISBN: 9781668440124
EISBN: 9781668440148
© 2022; 294 pp.
List Price: US$ **250**

ISBN: 9781799890782
EISBN: 9781799890805
© 2022; 306 pp.
List Price: US$ **260**

ISBN: 9781799868583
EISBN: 9781799868606
© 2022; 343 pp.
List Price: US$ **240**

ISBN: 9781799893158
EISBN: 9781799893172
© 2022; 261 pp.
List Price: US$ **250**

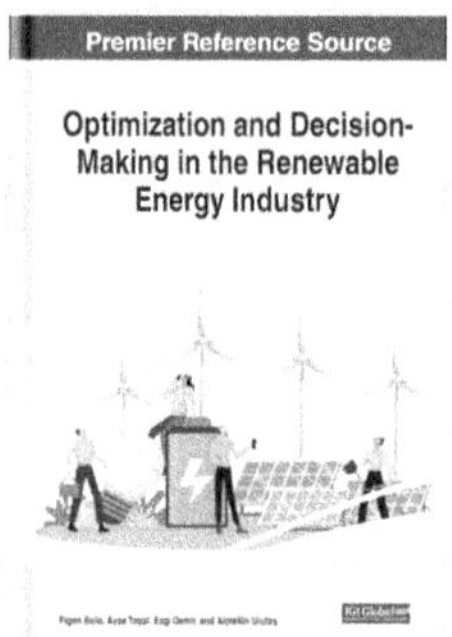

ISBN: 9781668424728
EISBN: 9781668424742
© 2022; 304 pp.
List Price: US$ **270**

ISBN: 9781799883746
EISBN: 9781799883760
© 2022; 366 pp.
List Price: US$ **215**

Do you want to stay current on the latest research trends, product announcements, news, and special offers?
Join IGI Global's mailing list to receive customized recommendations, exclusive discounts, and more.
Sign up at: **www.igi-global.com/newsletters.**

Publisher of Timely, Peer-Reviewed Inclusive Research Since 1988

IGI Global
PUBLISHER of TIMELY KNOWLEDGE

www.igi-global.com Sign up at www.igi-global.com/newsletters facebook.com/igiglobal twitter.com/igiglobal

Ensure Quality Research is Introduced to the Academic Community

Become an Evaluator for IGI Global Authored Book Projects

The overall success of an authored book project is dependent on quality and timely manuscript evaluations.

Applications and Inquiries may be sent to:
development@igi-global.com

Applicants must have a doctorate (or equivalent degree) as well as publishing, research, and reviewing experience. Authored Book Evaluators are appointed for one-year terms and are expected to complete at least three evaluations per term. Upon successful completion of this term, evaluators can be considered for an additional term.

If you have a colleague that may be interested in this opportunity, we encourage you to share this information with them.

Easily Identify, Acquire, and Utilize Published Peer-Reviewed Findings in Support of Your Current Research

IGI Global OnDemand

Purchase Individual IGI Global OnDemand Book Chapters and Journal Articles

For More Information:

www.igi-global.com/e-resources/ondemand/

Browse through 150,000+ Articles and Chapters!

Find specific research related to your current studies and projects that have been contributed by international researchers from prestigious institutions, including:

Massachusetts Institute of Technology — Harvard University — Columbia University in the City of New York — Australian National University

- Accurate and Advanced Search
- Affordably Acquire Research
- Instantly Access Your Content
- Benefit from the InfoSci Platform Features

"*It really provides an excellent entry into the research literature of the field. It presents a manageable number of highly relevant sources on topics of interest to a wide range of researchers. The sources are scholarly, but also accessible to 'practitioners'.*"

- Ms. Lisa Stimatz, MLS, University of North Carolina at Chapel Hill, USA

Interested in Additional Savings?

Subscribe to

IGI Global OnDemand *Plus*

Learn More

Acquire content from over 128,000+ research-focused book chapters and 33,000+ scholarly journal articles for as low as US$ 5 per article/chapter (original retail price for an article/chapter: US$ 37.50).

7,300+ E-BOOKS. ADVANCED RESEARCH. INCLUSIVE & AFFORDABLE.

IGI Global
PUBLISHER of TIMELY KNOWLEDGE

IGI Global e-Book Collection

- Flexible Purchasing Options (Perpetual, Subscription, EBA, etc.)
- Multi-Year Agreements with No Price Increases Guaranteed
- No Additional Charge for Multi-User Licensing
- No Maintenance, Hosting, or Archiving Fees
- Continually Enhanced & Innovated Accessibility Compliance Features (WCAG)

Handbook of Research on Digital Transformation, Industry Use Cases, and the Impact of Disruptive Technologies
ISBN: 9781799877127
EISBN: 9781799877141

Handbook of Research on New Investigations in Artificial Life, AI, and Machine Learning
ISBN: 9781799886860
EISBN: 9781799886877

Handbook of Research on Future of Work and Education
ISBN: 9781799882756
EISBN: 9781799882770

Research Anthology on Physical and Intellectual Disabilities in an Inclusive Society (4 Vols.)
ISBN: 9781668435427
EISBN: 9781668435434

Innovative Economic, Social, and Environmental Practices for Progressing Future Sustainability
ISBN: 9781799895909
EISBN: 9781799895923

Applied Guide for Event Study Research in Supply Chain Management
ISBN: 9781799889694
EISBN: 9781799889717

Mental Health and Wellness in Healthcare Workers
ISBN: 9781799888130
EISBN: 9781799888147

Clean Technologies and Sustainable Development in Civil Engineering
ISBN: 9781799898108
EISBN: 9781799898122

Request More Information, or Recommend the IGI Global e-Book Collection to Your Institution's Librarian

For More Information or to Request a Free Trial, Contact IGI Global's e-Collections Team: eresources@igi-global.com | 1-866-342-6657 ext. 100 | 717-533-8845 ext. 100

www.ingramcontent.com/pod-product-compliance
Lightning Source LLC
LaVergne TN
LVHW081323300926
849161LV00008B/1864